Advance praise for *Only*

"Science is done by real human beings, with human concerns. *Only the Longest Threads* tells a story that conveys the human side of science in a way that is as moving as it is accurate."
—Sean Carroll, theoretical physicist at Caltech
and author of *The Particle at the End of the Universe*

"Tasneem Zehra Husain writes lyrically, poetically about life, love, and physics. I highly recommend this wonderful book for anyone interested in what physics, and indeed all of science, is about. She masterfully describes the most momentous moments in physics history with verve and talent."
—Amir D. Aczel, author of *Present at the Creation*

"A delightful meditation on the development of modern physics, culminating in the discovery of the Higgs. Husain follows the thread of its creation through a dialog between a journalist and a young theory student, and as seen through the eyes of witnesses."
—John Huth, Donner Professor of Science,
Harvard University

"How do theoretical physicists think? Tasneem Zehra Husain knows. She knows their purpose, feels their passions, articulates their frustrations, shares their triumphs. Through the device of fiction *Only the Longest Threads* communicates the history of physical thought—its roots in inquisitiveness and essential disinterest in outcome—with greater clarity than any popular science text."
—Michael Duff FRS, Abdus Salam Professor
of Theoretical Physics, Imperial College London

"Well-written and cleverly constructed, this book takes us on a journey through the history of physics as a series of fictional adventures, loosely linked by another fiction, the storytellers' emails to each other. Some books are praised because 'I couldn't put it down,' but this one merits a deeper reading, one that stops, muses on, and savors each story before going on to the next. Each one captures not only the emergence of a significant idea in physics, but also something of the characters, culture, and times surrounding that development. So take your time, pause to ponder, but persevere—you will be well rewarded!"

—Helen R. Quinn, Physicist, Science Educator,
and co-author of *The Mystery of the Missing Antimatter*,
Professor Emeritus SLAC National Accelerator Laboratory

"Tasneem Zehra Husain's writing is both enlightening and entertaining as it captivates the challenge and excitement of working at the forefront of paradigm-shifting discoveries. Book-ended by the history-making discovery of the Higgs field, this tale offers a sparkling account of our understanding of fundamental physics. Through many voices rich with evocative metaphors, the threads woven through time and place that make up our current understanding of reality are revealed."

—Elizabeth F. McCormack,
Professor of Physics, Bryn Mawr College

"*Only the Longest Threads* describes the process of scientific discovery by focusing on the human elements: the bold conjectures, the wrong turns in the road, the competitiveness among scientists, the strength of their community, all seen from the point of view of the writers of letters and journals who make up the narrative. The clear, flawless prose is laced with a gentle wit when human behavior is described in the terms of physics—a welcome, light-handed nerdiness.

"Everyone who has studied physics but is unable to see the forest for the trees—and that means most of us—will relish this lovely little volume as it brings into perspective, through its accessible yet substantive treatment of the grand unifications, a magnificent edifice created by the human mind."

—Asad Abidi, Distinguished Chancellor's
Professor, UCLA Engineering

"Tasneem Zehra Husain skillfully weaves a poetic tapestry from tight threads of science and richly imagined strands of time. A weft of physics and warp of love makes a delightfully gripping read.

Her flowing prose conducts us by some unfamiliar force from falling apples to colliding protons where the Higgs boson looms."
—Joseph Mazur, author of *Enlightening Symbols*

"This highly original book puts a fresh perspective on humanity's inevitable obsession with understanding the laws of Nature. On an artfully constructed journey through space and time, Tasneem Zehra Husain gives us a tantalizing taste of how physicists struggle to find 'true nuggets of gold, and the only immortal elixir.'"
—Freddy Cachazo, Perimeter Institute for Theoretical Physics

"Tasneem Zehra Husain's observant narrators are witnesses to the intellectual revolutions of Newton, Maxwell, Einstein, Bohr, and others, the drama building to the mysteries of the present day. She uses her deep knowledge of physics to create a new genre—true science fiction, imagined vignettes of physics in all its humanity, woven together as a story within a story."
—Mark A. Peterson, author of *Galileo's Muse*

"*Only the Longest Threads* is a page-turner that portrays the excitement of discovery in physics from Isaac Newton to string theory. I highly recommend it for anyone who wants to feel the thrill of the succession of ideas that have led us to the current highly successful, but incomplete, understanding of our universe."
—Richard Dower, Chairman,
Science Department, Roxbury Latin School

"This book reveals a love affair, a love affair with physics. *Only the Longest Threads* is not about formulas and mathematics, but about people who have changed the way we comprehend the universe. We meet these icons, some ancient and some brand new, in a very personal way, and see that the driving force behind discoveries is very often a passionate relationship with the problems you try to solve."
—Olov Amelin, Director of the Nobel Museum

"With a self-referential structure reminiscent of Calvino's novels, and a premise—that fiction can bring physics to life—similar to McCormmach's *Night Thoughts of a Classical Physicist*, Tasneem

Zehra Husain takes us on a journey through epic discoveries as they might have been seen at the time by amateur enthusiasts. With the story framed by the discovery of the Higgs boson at CERN, and carried forward through e-mails between an aspiring young physicist and a struggling science journalist, the 'long threads' of the 'tapestry' woven by Newton, Maxwell, Einstein, are examined in the course of elegantly composed letters between imagined witnesses to physics history, until the narrative is taken out of their hands for a buildup to the Higgs boson discovery, and a nod to string theory."

—Paul Townsend, DAMTP Professor of
Theoretical Physics, University of Cambridge

"It is very rare to come across a book popularizing science that not only gives an excellent rendition of the development of a research area but also has high literary qualities. Tasneem Zehra Husain has managed to combine an outstanding description of several areas in physics while painting historical and geographical mini sketches, using voices from the times and places where key developments took place. Theoretical physicists will nod in recognition when reading about CERN, the Niels Bohr Institute, Trieste, Stockholm, Harvard and Cambridge University. The author is obviously sharing her experience of all these places. I have read many popular books describing various stages in the attempts to unify fundamental physics, but none as convincing and as good a read as this one."

—Ulf Lindström, Professor of Physics, Uppsala University

"Husain takes you by the hand and walks you through half a dozen of the most important discoveries in physics over the past three centuries, in each case as seen through the eyes of a fictional protagonist living through the discovery. Along the way she gives you an almost personal sense of how science feels as it is happening, along with distilled insights into the essence of the discoveries themselves. These vignettes are wrapped in a story that is in itself a page-turner. I can't wait for the sequel."

—Krishna Rajagopal, Professor and Associate
Department Head for Education, Physics, MIT

ONLY
THE
LONGEST
THREADS

Tasneem Zehra Husain

PAUL DRY BOOKS
Philadelphia 2014

First Paul Dry Books Edition, 2014

Paul Dry Books, Inc.
Philadelphia, Pennsylvania
www.pauldrybooks.com

The story "Quefithe" by Predrag Cvitanović on pages 185–86
is reprinted from *Field Theory*, http://ChaosBooks.org/FieldTheory,
Niels Bohr Institute (Copenhagen 2004).

Printed in the United States of America

Library of Congress Cataloging-in-Publication Data
Husain, Tasneem Zehra.
  Only the longest threads / Tasneem Zehra Husain.
    pages cm
  ISBN 978-1-58988-088-7 (alk. paper)
  I. Title.
  PS3608.U795O56 2014
  813'.6—dc23
                          2014024996

Contents

For Ammi and Abbu,
Because all my stories start with them

ONLY THE LONGEST THREADS

*Nature uses only the longest threads
to weave her patterns, so each small piece
of her fabric reveals the organization
of the entire tapestry.*

—RICHARD FEYNMAN

A Poem in the Mind

If you think for yourselves, as I proceed, the facts will form a poem in your minds. —MICHAEL FARADAY

5:00 A.M., JULY 4, 2012
CERN, GENEVA

Sara

Between sleep and wakefulness lies a moment of indecision. For that vague, foggy instant, the act of being is diffuse and ill defined. You seem to inhabit a superposition of states, to be dispersed between multiple existences. It passes quickly. The wave-function collapses. Consciousness flows into your own reality, having chosen, from many possibilities, this particular one. I don't always applaud my selection when I am fully awake, but I have done well today. This morning, there's nowhere else I'd rather be than right here, part of the historic Higgs boson stake-out at CERN.

From the moment this seminar was announced, rumors started to fly that the elusive particle had finally been sighted. I knew this was my one big chance to watch conjecture turn into fact, so I wrote coaxing emails, made pleading phone calls, wangled a visitor's pass, and bought my train ticket to Geneva. The Higgs boson has been eagerly anticipated for decades, but the instant that possibility turns into certainty, our knowledge will undergo a phase transition; it will be a turning point, a step

1

function, or however else you care to describe a discontinuous change. Whenever I read about discoveries that changed the landscape of science, I wonder what it must have been like for people who were there at the time, who felt the ground shake under their feet. Today, if all goes according to plan, I may experience that jolt myself. If the predictions are borne out, we could, in a matter of hours, be suddenly and irrevocably ushered into a new era.

I hear the hum of voices even before I raise my head. The doors don't open till 7:30 a.m., but I had no intention of missing my date with destiny, so I arrived shortly after midnight. There were just a few people here, some huddled over laptops, others reading. Between then and now, the crowd has grown exponentially. Under my gaze, scattered groups spontaneously organize themselves into a pattern. Like data points added to a plot, they collect in clumps, then spread outward to swallow up the empty spaces and merge into a continuum.

The line snakes around from the entrance, past the council chambers, and down the stairs. Those of us who got here early enough to rest are now back on our feet. Many have that unmistakable graduate-student look. I would recognize it anywhere, this mark of our tribe. Some head to the restrooms to splash water on their faces, others depend on coffee to wake them up. As if on cue, the shrill ring of a fire alarm resounds through the halls, but no one moves. I guess we'd all rather go down in smoke than risk losing our place in the line.

The crowd approaches rock concert proportions as reporters, photographers, and journalists join us in line. It strikes me there's a Higgs boson paparazzi even before we have proof that the boson exists, and I laugh out loud. A few heads turn at the sound, but one smile beams through all the cautious glances. It's a friendly smile, knowing, even a little sympathetic. The guy who belongs to it has dark hair, dark eyes, and looks vaguely familiar. This is not just wishful thinking. I've definitely seen him before, but I can't remember where.

He's talking to someone on the phone. Is that an Italian accent? Before I can decide for sure, the doors to the auditorium open. A wave flows through the crowd, and in moments I am swept—along with hundreds of others—into the inner sanctum.

7:40 A.M., JULY 4, 2012
CERN, GENEVA

Leo

Where did she go? She was right here jostling to get into the Main Hall a minute ago. I would never have lost sight of her if Joe hadn't called. It's not like he even had anything sensible to say. "Everyone and their brother will be writing about this boson today, Leo. I'm counting on your piece to stand out. Make it sound new." Might as well make spring sound new. Or love. Or anything else that's been written to death.

I don't usually let Joe get to me; his words rankle only because he touched a nerve. I should never have confessed my increasing dissatisfaction with journalistic pieces. I still want to write about science, I told him, but in a different way. When Joe asked, as any editor would, what this new way was to be, I had no coherent answer. All I could say was that even when my exposition is clear, I'm aware there's something missing. I can sense the fissures between words, but I can't figure out how to fill them in. There are times when I feel the answer hover just beyond my grasp, but when I reach out, it vanishes into vapor. The damn thing is like Schrödinger's cat, I said. I kill it by looking at it.

Somewhat bemused at that analogy, Joe told me he had no issue with a fresh angle, but that until I found it, I still had articles to write. "Don't spend all your time experimenting with new voices," he warned. "We work at a magazine, not a literary journal." Having said that, how could he then turn around and remind me of both the time crunch and my rankling desire for an original approach? Annoyance at Joe's ill-timed call distracted me just long enough to lose sight of the girl with the red backpack. The fact that I have not yet figured out how to pitch today's piece further fuels my irritation. Enough time has been wasted trying to approach the problem from an oblique angle; now it must be faced head-on.

Before Joe disturbed my equilibrium, I was reasonably content. Amused and excited by the spectacle here, I hoped something I saw or heard would trigger the flow of words. As I glanced around, looking for a hook on which to hang my article, I was struck by the number of people glued to their devices, surfing the net on laptops, tablets, and smartphones. Do they realize this is the birthplace of the World Wide Web? I wondered. Does anyone think about the fact that here, where the digital universe originated, the origin of the physical universe is also being probed?

Unlike paleontologists and archeologists, physicists don't have ruins or fossils to work from. Nothing has passed down to the ages unchanged; it has all morphed, evolved, coalesced. The truth cannot simply be excavated, it must be recreated. When the universe was new and ablaze with energy, several particles bubbled up only to burst. Between their emergence from chaos and their immediate decay into oblivion, these particles were privy to many secrets. We believe the Higgs boson formed a part of this league, that it will have much to tell us, but we can't know for sure until we actually meet it. The only possible way to do that is to generate, again, the primordial energies that first brought this boson into being. Such is the task of the Large Hadron Collider, which lies buried in the ground under my feet. Beams of protons run frantic laps around a giant ring, building up speed until they attain energies that have not been seen since the first few seconds after the Big Bang. Then they are made to crash head on, so that their energy is unleashed and reshuffled into a plethora of other particles. One of these, we hope, will be the elusive Higgs boson.

Could that be a possible start to the article? I ran through the words again in my mind, but I couldn't figure out how to extend the "origin of the universe" argument or how to build on the digital-physical analogy. That's when I heard it. A laugh as melodious and lilting as a caged lark set free. I saw her as soon as I turned around, a rueful smile on her face—as if the laugh had escaped accidentally. I smiled back, in what I hope was a reassuring manner. And as I looked at her face, that soft brown hair, those hazel eyes, I knew I had seen her before.

Last spring, I did an in-depth piece on the High Energy Theory Group at Harvard University, featuring profiles of several professors and discussions of their research. I visited the campus to conduct these interviews, and when I reached the red brick building that I thought was Jefferson Laboratory, I saw a girl sitting under a pink tree, petals scattered around her as she bent over her books, chewing on a pen. I had a sudden urge to confirm my whereabouts, so I went up to her and asked where the Physics Department was. She pointed straight ahead. I told her whom I was looking for, and she explained how to get to the right office. A few words, a few smiles, and I was on my way.

That was our only interaction, but today I remember every detail distinctly. Transient objects can have a lasting impact, I caught myself thinking. Ephemeral particles that flash in and out of existence play a crucial role in the way our universe unfolds. I stopped that thought in its tracks. I need to write an article, not a sentimental greeting card. Still, I craned my neck to look for her.

Through the pulsating gaps that formed as people moved restlessly back and forth, I caught a glimpse of her backpack, and then it disappeared again. Just like that, a phrase came to me: "Interactions between elementary particles are shielded from prying human eyes by a thick fog of data." Now there's a possible opening line! My brain pounced on the sentence. I told myself there would be time enough for writing later—after I had found the girl with the shining smile—but words have a will of their own and the article continued to unfold in my mind.

"Those of us who would learn the secrets of the subatomic realm must sit patiently, poking holes in the mist, waving it aside, hoping for it to part. This morning the fog cleared for an instant and we caught a glimpse of a new particle that is, most probably, the long-sought Higgs boson." Hmm. Not perfect, but not too bad. And way too long to remember. I reached for my pen to jot the sentences down before I forgot them, and just then, without warning, Joe called, the doors opened, and I lost her in the commotion.

Settled in the press room, I fix my gaze on the large screen that hangs right in front of me, all set to broadcast the historic proceedings. As I peer through this virtual window into the Main Hall, I see people sitting on the stairs. There is not even standing room anymore. I scan the rows for the girl with the smiling face, but I can't make her out in the crowd. Maybe she is just my muse. She did inspire my initial lines after all. The piece last year went really well, I remember. Seeing her again might be a good omen for this article, too. But even as I tell myself that, my eyes are focused on the screen, trying to lock in a red backpack. No luck so far. I wonder if experimentalists feel similarly frustrated when they scan reams of data, looking for a blip, an anomaly—something that stands out . . .

Sara

Was that a smile of recognition or just a friendly response to my random laughter? Have we met before? As I look for a place to sit, I thumb through the images in my mind. He's obviously not someone I went to school with. He looks Italian, or thereabouts. Where, and why, would I have met him? I tuck my knapsack under my seat, and as I bend back up again, I know.

He's the reporter who did that story on the Theory Group last year. We met outside Jefferson and I directed him to the secretary's office. All of us at the Physics Department loved how the article came out. It was balanced and well written. He didn't make caricatures out of the researchers or gloss over their science. But what made it sensational was the way he captured the spirit of the work, of that life, and the people who live it. The article was pinned up on the board for the longest time, and for months, as I passed through the corridor, I saw a little photograph of him, together with the byline. His name is Leonardo . . . something. When I told my friends about our little run-in, they did not mince their words. "He's cute. He gets physics. He expresses himself so well. How could it be better? Why were you in such a rush to send him off? You should have found something to say!" If they knew I had seen him again here, I would be subjected to a tirade on how opportunities are wasted on the undeserving, and then treated to a display of spontaneous internal combustion as my friends went up in smoke.

I have barely settled into my seat when there is a tap on my shoulder. A girl in the row behind me wants to know if I can take a picture of her and her friends. I take three or four, just to make sure no one's eyes are closed. All around me, cameras—and camera phones—click as people create tangible records of something as intangible as a moment in time. Our instinct to capture the moment is so strong. Throughout history, we have recorded our travels through paintings and photographs. On the journeys of the mind that we now undertake, abstract data is often the only souvenir we can bring back. So we turn it into plots and graphs, looking for ways to illustrate relationships and present information visually. Anything to make the alien familiar.

The air in this room is thick with a heady mix of anticipation and excitement. I wish I could capture its essence somehow. I am reminded of the souvenir shops I saw a few days ago, near Place de la Concorde: rows of charming glass vials with extravagant labels that read "L'air de Paris." Outwardly I scoffed at the gullible tourists who fell for this sentimental trap, but secretly I was tempted to buy one of the pretty bottles for myself. I imagined opening a vial at home and letting the air of Paris fill my apartment. The joy of uttering that last phrase would almost justify the exorbitant price of an empty bottle. If my friends hadn't dragged me away exactly when they did, I would have given in and bought one.

The desire to possess an ambience is even stronger now. This, right here, is one of those rare occasions when the thrill of theoretical physics is palpable. Most of our life is spent dealing with ideas that are abstract and cannot be embodied; they can only be held in the mind. Even when it feels like we have moved mountains, there is nothing physical we can point to as a sign of our accomplishment.

The tools of our trade are also intangible. We have nothing material to clutch or discard, no object that would absorb or reflect our emotions. Theoretical physics is largely a private affair, a life lived out in the mind. But here and now, the passions that rule us seem to have flowed out and found an external manifestation. I'd take anything, even a meaningless prop, that would help me evoke the atmosphere of this auditorium again. If they were to start selling ugly empty tin cans labeled "L'air de CERN: 4 July 2012," I would stand in line to buy one.

Then, when people ask me what I study, and why, with all the concrete problems left to solve in this world, I spend my days worrying about abstractions, I could hand them this can. Sniff the profound joy of a mystery solved, I would tell them. Take a deep breath and inhale the triumph. This air is saturated with the best of what the human race is capable of. Curiosity. Inspiration. Creativity. Collaboration. Ingenuity. Persistence. Purpose. Feel the thrill of recognition that runs through you. This quest for knowledge is what elevates us. It gives us honor and dignity. This is what we were made to do.

But I couldn't give anyone that flowery speech without convincing them that I was mad! There has to be a less fanciful way to communicate the lure of fundamental physics. I could recite a list of all the tangible benefits that have accrued from theoretical research—all the way from electricity to GPS and PET scans— but that seems a bit defensive to me. Truth be told, people pursue pure research because it satisfies a basic human need. The urge to ask questions and seek answers cannot be denied without stifling the spirit. How does one convey that elemental sense of wonder and fulfillment to people who lack technical expertise? Without going into the nuts and bolts of an idea, how do you get someone to feel its tug?

There is a beauty and a strength in our constructs, and if the Higgs boson is revealed today—as the rumor mill predicts—the truth of that statement will become manifest yet again. Thanks to the unprecedented media coverage of this event, practically everyone has heard of the elusive boson that has been chased for decades. A lot of ingenuity, innovation, and diligence have gone into hunting down this "last missing piece of the Standard Model." But the Higgs boson is far more than just an addition to the particle physics roster. It is a symbol, a testament to the power of arguments that we're able to sense the absence of a particle whose existence left no trace. If found, the Higgs boson vindicates not only the Standard Model of particle physics, but an entire system of thought; it is proof that our equations have a wisdom and logic all their own, that they know more than we have put into them.

Tumultuous applause breaks out as the big guns walk in and take their seats. I recognize Peter Higgs, of course, and also Fabiola Gianotti and Joe Incandela, the spokespeople for ATLAS and CMS, the two detectors that were independently chasing the Higgs. It will be interesting to see how their results matched up. CERN's Director, Rolf-Dieter Heuer, takes the podium.

In minutes, the screen in front of me will be flashing images of machinery, plots, results from the world's most sophisticated experiment; but right now, the only projection is a sea of faces, in an auditorium, much like this one, at the other end of the world. At the international particle physics conference in Melbourne, physicists are glued to their chairs, in anticipation of what is to come.

"Today is a special day," begins Heuer. It's show time.

Leo

Two hours before my train departs. I want to put the finishing touches on my article and send it off to Joe so I can catch some sleep on the train back to Milan. I settle down on a bench and pull out my laptop, and I read through what I have. At the end I add this:

"Today's results from the ATLAS and CMS detectors, though preliminary, were in stunning agreement. Both experiments independently found conclusive evidence of a new particle in the 125–126 GeV energy range. On the surface this boson, the heaviest found to date, has every appearance of being the missing piece that completes the Standard Model, but scientists must ensure that it fulfills the entire set of criteria before they officially refer to it as the Higgs boson. Whatever properties it turns out to have, this particle is definitely the first of its kind. Neither matter nor force, it belongs to a whole new category entirely. 'We have reached a milestone in our understanding of nature,' said CERN Director General Rolf-Dieter Heuer. More detailed studies of the properties of this new particle are 'likely to shed light on other mysteries of our universe.'"

Not quite the piece of art I was holding out for, but I have a deadline. This will have to do, for now.

Even so, I can't help fidgeting. I could never resist peeling a scab either. I quite like the beginning, I decide, but the article sags in the middle and the end just falls flat. I tweak a few words, change them around. Still nothing. It's clear and concise, but it doesn't . . . sing. If I could pinpoint the problem, I might be able to fix it. I read it one more time, slowly. The facts are all there, but the euphoria doesn't make it through. Why? I've said this boson is the first of its kind. I've called the discovery a milestone. I've even alluded to the mysteries of the universe. What else is one to do? What did I leave out?

Underneath my growing irritation, my mind grapples with this question the only way it knows how. Words float up to the surface. Fractured phrases struggle to express the problem, to give it shape and dimension. Nothing sticks until a memory of the natural history museum flashes into view. Why this? What's the message here? I describe the image to myself, trying to work out what it means. And then it hits me. My last paragraph feels like an exhibit I once saw: an entire wall of butterflies, neatly pinned down in display cases. Textbook perfection, but it left me cold. No matter how long you stare at its frozen wings, you can't really know a butterfly until you witness the chaotic abandon of its unchoreographed flight. That exhibit was too carefully arranged. There was no movement, no life, no opportunity for accidental encounters. You could smell the chloroform in that room. The ending to my article has a similar odor.

Diagnosis is a relief, but it brings me no closer to a cure. It's all very well to say the writing needs movement and life, and room for serendipity, but how do you make that happen? I can feel a headache coming on. A stroll around the station might be a good idea. I notice a café about ten steps away. I can't remember the last time I ate. Maybe food will do the trick.

As I weave through the tables to the counter, my foot catches in something, and I trip. When I see suspiciously familiar red straps entangled around my shoe, I glance up, but all I can make out from this angle are a pair of funky glasses. Soft brown hair slips out of some clip-like contraption and falls like a curtain across her face. As I pull my feet out of the straps, she turns with a start. Yup. It's definitely her.

"Oh, I'm so sorry," she says. "I thought I'd put it away under the table. Are you okay?"

"I'm fine," comes my rushed assurance. For a second, I am torn as to what to do next. Talk about lucky accidents. She's here alone. The situation is tailor made for conversation, but I've been caught off guard. I don't want to let the opportunity go, but I don't want to appear too pushy either. What do I say? As I cast around for casual, friendly opening lines, the Higgs boson saves the day. In this moment, science enthusiasts across the world share the camaraderie of sports fans whose team has won a major tournament; the usual social barriers between strangers are bridged by shared joy. Without letting myself overthink this, I say, "Excuse me for asking, but weren't you at CERN this morning? I thought I might have seen—"

"Yes, I was," she replies, the excitement in her voice evident.

"Do you work there?"

"No, no," she says. "I'm just here as a tourist, really. I was visiting Paris when I heard about the seminar and, being so close by, I couldn't resist coming down." There's a slight pause. "I'm a graduate student," she adds, as if the earlier sentence was incomplete. As if her enthusiasm needs to be justified. "I study physics."

"You came to Geneva just for this?"

"Yes," she nods.

Just the one word, but there's a hint of pride in it. Something about her arched eyebrows and widening smile makes me want to know more. This, right here, is the reaction I want my readers to have.

"I'm a writer," I say, "based in Milan. I came up to cover the announcement this morning."

"That's really cool."

It's still tentative, and maybe just polite interest, but I'll take it. "Most of the time, I would agree with that," I say, "but I'm a bit stuck right now. Actually, would you mind if I asked you a few questions? Your perspective could be really helpful for something I'm working on."

"Sure," she gestures to the empty chair at her table.

This is going even better than I expected. For a moment, I contemplate mentioning our brief encounter last year. Would that put her at ease, or would it seem creepy that after all this time I remember her? She gives no sign of having recognized me, so I think I'll let it go. I catch myself drumming my fingers against the tabletop. Stupid nervous habit. I think I stopped before she noticed.

"So, I should explain. I write about science for a general audience, but a lot of the time, I'm not too happy with the way the tone comes out. I want it to sound less journalistic and more alive. Something engaging, that will draw readers in, instead of shutting them out, you know? So I'm trying to figure out what sorts of things make people excited about science."

Another nod. Another smile. But nothing else.

"What exactly do you study?" I ask, before she writes me off as a self-obsessed egomaniacal wound-up writer type.

"I'm aiming for string theory."

"I've heard of your kind! You're the ones who move around in ten dimensions." For one everlasting instant, there's silence. Your turn now, I think. Talk. Say anything. Or is the joke too much, coming from a stranger? And then she speaks.

"Guilty," she says. "But on days like today, we come down to four."

This conversation is picking up. I can't let the momentum drop now. "So, going back to the excitement of this morning. The event was being live-streamed and news was updated constantly. You would have found out what happened, pretty much instantly, regardless of where you were," I say. "What made you want to come here in person?"

She crinkles her nose and looks up at the ceiling. "Well," she says, "this may sound a little silly, but the Higgs boson is something I've read about my entire student life. Most of the people at CERN this morning have taken exams that involve this particle. Our grades have depended on it, yet no one knows for sure if it really exists. And then, out of the blue, I find out that today it could be walking off those textbook pages and into the world. It's like a fictional character coming to life, you know? I wanted to be here to greet it."

A loud announcement over the speaker system snaps me back to a sense of my surroundings. We're sitting at a train station, headed to different cities. This serendipitous encounter is bound to end soon, and I'd like to know how much time I have left.

"When does your train leave?" I ask.

She glances at her watch. "I have about an hour," she says. "What about you?"

A little over an hour, I tell her. Not nearly enough time. In case she quantum-tunnels out of my life, disappearing as unexpectedly as she materialized, I want to be able to find her again. The least I should know is her name. I kicked myself for not doing this last year. I'm not making the same mistake again.

"Oh, by the way, I never introduced myself properly. I'm Leo," I say.

"Nice to meet you, Leo. I'm Sara."

Sara

"I'm Leo," he says.

I have to bite my lip to keep from saying "I know." As I go through the "nice to meet you" routine, I remind myself not to give away the fact that we've met before, if you could even call that a meeting. He obviously doesn't remember, but then my photograph didn't stare at him for months, every time he stopped to check what was new in his department, or division, or whatever you call the place he works at.

"You were saying," he prods?

"I guess I just wanted to be there in the moment," I say. If he wants to know more, he can ask.

"And was it everything you thought it would be?" His tone is genuinely curious, as if he's trying to puzzle this out.

"Absolutely!"

Leo looks at me expectantly, waiting for me to continue, but I don't know how to answer. Once I get started on this topic, I have a tendency to run on. I'm not sure I want to get that deep into conversation with someone who will probably forget what I said. On the other hand, he's being quite nice and there's no reason to brush him off. It's true that he doesn't remember me, but I almost caused him a fall. I'd say we're even. And anyway, answering this particular question isn't a bad idea, even for self-ish reasons. He writes well, and if he figures out a way to inject the atmosphere of this morning into his article, I can just pass that on to friends and family. It'll save me endless, repetitive ex-planations of what the Higgs boson is, what it does, and what that means. So, in the interest of good science writing, I decide to give him a proper answer.

"How do I explain? We all expected the Higgs boson to show up one day. This isn't something we need to shift around and make room for; we've been saving it a seat at the table for years. Even though the discovery is monumental, this particle has al-ready been incorporated into our scheme of thought. It will not cause particle physics texts to be rewritten. All that will hap-pen is that the customary reference to the ongoing search will change tense. In future editions, a new sentence will be added, to say that the Higgs boson was discovered at CERN in 2012. In a few years, to a new generation, that is all it will be: a flat, black-and-white statement of fact. But for those who have lived through this moment, that same prosaic phrase will sparkle and gleam. To us, that bland sentence will never be an end, but a point of takeoff for wonderful memories."

That sounds borderline insane, but he's watching me intently with his dark eyes as if he understands.

"Like a secret message, hidden in dry prose," he says. "Now that you have the key, you can unlock the meaning."

Exactly! He gets it.

"It's fabulous," I say, "but at the same time it makes me won-der what lies behind all the other doors that are still closed to me. So many other sentences in our textbooks contain these glimmering messages about great discoveries of the past, but I don't know how to unlock them . . ."

"I suppose you just have to be there when they hand out the keys," says Leo with a rueful smile.

"I guess so," I admit reluctantly. But as the words leave my lips, I realize there is another way.

"When I was about seven years old, I discovered a photograph of my grandmother at the same age. I was spellbound by the girl in the picture. If we had met, I would want her to be my friend. She was a slight, wiry thing who stood up straight and gazed into the camera with a naughty defiance. My grandmother had a slight stoop, skin that fell in soft folds, and kind eyes. When I took the photograph to her, she lit up, and for a second I glimpsed the little girl inside. I spent the afternoon listening to tales of her childhood and youth. I began to see me in her, and her in me. From that day on, our relationship changed."

Another sprawling speech. It appears Leo kept pace, but will he see the point? He does.

"So you did become friends with the little girl after all," he smiles. He pauses for a moment, then continues more slowly. "I see what you're saying. Watching something, or someone, in their infancy is an act of privilege. It grants you a degree of intimacy."

Yes, he's right. When things are new and vulnerable, they tend to draw you in. It is easier to bond with the unformed than with the mature. But there's more to it. That photograph gave my grandmother dimension. It made me aware that she existed in planes other than the one where we overlapped, that her reality was deeper than what I knew. Even as I pestered her for stories, I realized I would never learn everything about her, never know all that she had been. Under the sepia glow of that long-ago image, my beloved, comfortable grandmother had suddenly been tinged with mystery.

As I sit, lost in thought, Leo too is silent. He speaks now, but his brows are still furrowed.

"This time travel thing could be just the trick, actually. For a while now, I've wanted to write a history of sorts, about theories that changed our understanding of the universe—like gravity, for instance, or relativity. But every time I try, it comes out too wooden. Even when the ideas are crucial, it's hard to get excited about things that happened so long ago. Looking back a few centuries, hindsight casts its dull light on work I know to be brilliant. Maybe the answer is in your grandmother's picture. Maybe I need to go back to a time when these theories were new."

"That sounds like a pretty good idea. It would be fun to immerse yourself in the past and look at familiar theories through eyes that do not yet know what is to come . . ."

Leo turns away slightly and stares intently into space, as if trying hard to bring something into focus. His nose scrunched up, he shakes his head. "But that can't be all. If being in the moment was all it took, my article about this morning should be bursting with life. And it's not. I'm quite happy with the way it starts, but the end is driving me crazy. I've tried to capture all the key details of the day. I think it paints a decent picture. The facts are all there, it's just that . . ."

"Facts aren't science," I say, before I cut myself short. I've quoted that phrase so many times, it has become a reflex action.

Leo raises his eyebrows, motioning for me to continue. I have to finish the thought.

"'Facts are not science, as the dictionary is not literature.'* I read that somewhere years ago, and it still echoes in my mind."

"That's it!" Leo says, his excitement evident. He starts talking, but at this point he's just thinking aloud, so I get out of the way and let his train of thought speed ahead. "It's the connections that create meaning. Literature lies in the way you link words, and science in the way you link facts. The shape of a story or a theory depends on how we join the dots."

*Martin H. Fischer, *Fischerisms*, ed. and comp. Howard Douglas Fabing and Ray Marr (Springfield, IL: Charles C. Thomas, 1944).

But you know that already, I want to say. You've joined those dots before! I think back to last year's article. It worked so well because he wove the researchers and their research inextricably together. The reader got a sense of the relationship between the people and their ideas, and a glimpse of the journey they were on. I can't quote that example without giving myself away, but I can talk about the piece he's working on now.

"The reason I came here today had nothing to do with the facts. Those, I could have found out from anywhere. I came to experience the emotions that were running rampant in that room. I came to drink in the atmosphere."

I can feel the weight of his entire attention on me.

"This article about the Higgs boson, the one you said you weren't too happy with, did you write about the feelings that erupted here this morning? Did you capture the jubilation of all those whose lives have been affected by, or even dictated by, this search? Did you talk about the excitement, the confusion, the blood, sweat, and tears that have gone into this search?"

He shakes his head slowly. No.

Well, then, how did you expect it to come alive? I want to say. But I don't know him well enough for that, so I dole out a more measured and appropriate response. "I guess what I'm saying is, you don't need to tell your readers every last fact; just show them how to create meaning. Show them the connections. Show them the science!"

"Show, don't tell?" asks Leo with a quizzical grin. "Isn't that the hallmark of fiction?"

The moment he utters those words, I know we've hit upon something big. I fight the urge to check my watch, but I can feel time slip away.

"So what's wrong with fiction?" I counter. "You said you were looking for a new approach, right?"

His thoughts are so loud, I can practically hear them. *Not that new*, they shout. But from what I've read of his writing, I know he can do this. I just hope I can convince him before my train leaves.

Leo

Fiction? The thought terrifies and fascinates me at the same time. I don't do fiction. Never have. But maybe that's the problem. Maybe Sara's right. All those qualities my writing lacks are precisely what fiction excels in. Already I can see that if the fissures in my prose were filled with emotion, the fractured segments would hold together as stronger, more coherent pieces. But even if fiction could solve all my problems, who will solve the problem of writing fiction? I've never done anything like this before. How would I even begin? I can feel the synapses fire in my head as my mind tries to work this out, but I look away.

"That's a great idea," I say, "for someone else."

Sara looks exasperated, as if I've frozen a roller coaster midride and brought the fun to a screeching halt. "Why not you?" she asks.

"I don't know how it would work. To me, fiction means people, and I don't really write about people. I want to write about ideas. Transformative ideas . . . ," I say, hoping that word will soften the blow. Sara does not respond, so I continue, ". . . theories that changed the course of physics."

That provokes a smile, and a response. "I thought you said you wanted to write about theories that changed our understanding of the universe."

"It's the same thing," I say, happy to have her involved in the conversation again.

"No," she says. "If our understanding changes, we're involved. People are a part of this story automatically, whether you like it or not. Think about it for a minute," she says. "Science is something we construct in tandem with the universe. It's a conversation between nature and us. If you ignore us, you leave out half the story. You want to write about ideas? Fine, write about ideas, but do it through people. Tell the story of physics, the way it is experienced."

Sara's eyes burst with light. Faced with her conviction, my resistance begins to falter. Could fiction really be an answer? Is this the "new style" I was looking for?

I think she can tell I am wavering, so she moves in for the kill. "Ideas don't exist in a vacuum, you know. They materialize in our minds. That's what makes it so hard for theoretical physicists to share our work with the public. We seldom have anything material to show, and we can't very well invite people into our minds, where all the action unfolds." Sara takes a quick glance at her watch. "But *you* can. You can take your readers and put them into someone's head, because that's what fiction does."

With that last sentence, a riot breaks out in my brain. She's right. You can always recognize your thoughts when you hear them, even if you couldn't have said them yourself. In some ways, that's the joy of literature; when we read, we encounter our own thoughts on the page. Ideas that were jumbled up in our minds, emotions that struggled for expression, find release. And all that pent-up tension dissipates.

Sometimes we think we don't understand things just because we've never heard them said in a tone that resonates with us. What if there are people out there who think they don't understand physics just because they haven't heard it expressed in words that strike the right note? What if, instead of being handed glib, polished statements, they could overhear the thoughts of a scientist who is processing new ideas and struggling to express them? Maybe these people would see something familiar, maybe they would recognize some of these tangled thoughts as their own.

And if that isn't a reason to try fiction, I don't know what is.

Sara raises her eyebrows in anticipation. "Well?" she gestures. A stray voice in my head says, "She speaks with her hands? How Italian." Out loud, I'm more cautious. "I guess I could give it a shot," I say. Daunting, but worth thinking about.

"Oh, please!" Sara says. "It's a brilliant idea, and I know you can do it." Before I can ask how or why she knows this, she pulls her backpack out from under her chair and starts putting her things away. We have only minutes left.

If we end on this note, we have no way of, or at least no excuse for, keeping in touch. I could just come out and ask for her number or email address, but despite the sparkling connection we have shared, and even though I can apparently pour the contents of my head out with reckless abandon, it's way too soon to take the lid off my emotions.

"I know you have a train to catch, so I won't keep you. Thanks for all these wonderful ideas. It's been so much fun talking with you."

"It was a pleasure. I look forward to reading your book."

That sounds too much like a final goodbye. I think fast. The best I can come up with is this: "I don't know about the book yet, but my slightly rewritten article about the Higgs boson should come out very soon . . ."

I keep my fingers crossed that she will follow up as I think she will. She does.

"I'll keep an eye open."

"If you want, I can email you the link when it's live."

"That would be great." Sara tears a little piece of paper from her notebook, scribbles something down, folds it up, and hands it to me. "I really need to run now." She zips up her backpack. "Bye."

As she walks away, I unfold the paper in my hand. Breaking.symmetries@gmail.com, it reads. I can't hold back the grin that spreads across my face. That's how all the best stories begin, I think. With the breaking of symmetry.

"Enjoy the rest of your trip!" I call after her.

Sara turns around and waves, and then heads toward the tracks. I keep watching that red backpack as it slowly grows smaller.

Email: Leo and Sara

From: Sara Byrne <breaking.symmetries@gmail.com>
Date: Fri, July 6, 2012 at 3:41 PM
Subject: Hi
To: Leonardo.Santorini@gmail.com

Hi Leo,

Thanks so much for sharing your article. I loved it. But now that I've seen how well you write, I'm even more convinced you should do the book. I thought about our conversation on the train back to Paris. Much as I love reading popular science, I don't think there's enough out there that conveys the passion of living a life surrounded by ideas. Your book could provide wonderful new insights to a general audience, so please do think about it seriously. I hereby volunteer to be a reader and a sounding board if you need one.

All the best,
Sara

From: Leonardo Santorini <leonardo.santorini@gmail.com>
Date: Sun, July 8, 2012 at 11:21 AM
Subject: The Book
To: breaking.symmetries@gmail.com

Hi Sara,

I'm so glad you enjoyed the article. Thank you for your offer of help; I might just take you up on it. I have to admit I can't get this book idea out of my head either. In fact, in some strange way, I feel like I've been moving toward it for a while . . .

But since this isn't exactly a traditional approach, it would help a lot if I could talk things over with someone during the writing process. Would you be willing to take a look at the chapters as they're done, and give me your feedback? If you're too busy, I will of course completely understand.

Cheers,
Leo

From: Sara Byrne <breaking.symmetries@gmail.com>
Date: Fri, July 13, 2012 at 1:24 PM
Subject: I'm in.
To: Leonardo.Santorini@gmail.com

Leo, we have a deal.

Consider me signed up.

Sara

From: Leonardo Santorini <leonardo.santorini@gmail.com>
Date: Sat, July 14, 2012 at 9:04 PM
Subject: Thank you!
To: breaking.symmetries@gmail.com

Sara, that's wonderful. Thank you.

I'm outlining the book now, and while I've managed to create narrators to walk us through each chapter in the evolution of physics, I can't help thinking how completely unnecessary it is to "make up" a string theorist? Won't you write the last chapter?

Ciao,
Leo

Email: Leo and Sara

From: Sara Byrne <breaking.symmetries@gmail.com>
Date: Sun, July 15, 2012 at 11:15AM
Subject: Quit Stalling!
To: Leonardo.Santorini@gmail.com

Leo,

We'll cross that bridge when we come to it!

For now, I'm waiting to read the first chapter.

Sara

From: Leonardo Santorini <leonardo.santorini@gmail.com>
Date: Thu, Oct 25, 2012 at 1:53 PM
Subject: Checking
To: breaking.symmetries@gmail.com

Hi Sara,

Hope this message finds you well. It's been a while since I last checked in with you, but writing took so much longer than I thought it would. It always does! I'm finally done with the first couple of chapters, but before I send them to you, I wanted to make sure you still have the time to do this. I will, of course, greatly appreciate your feedback, but I don't want to impose . . .

All the best,
Leo

From: Sara Byrne <breaking.symmetries@gmail.com>
Date: Fri, Oct 26, 2012 at 07:39AM
Subject: Waiting
To: Leonardo.Santorini@gmail.com

Dear Leo,

Of course I still have the time to do this. In fact, I've been waiting. Send it on over.

Sara

The First Installment

From: Leonardo Santorini <leonardo.santorini@gmail.com>
Date: Mon, Oct 29, 2012 at 11:21 PM
Subject: The First Installment
To: breaking.symmetries@gmail.com

Hi Sara,

Here are the first two chapters. Writing can be a lonely process at the best of times, and given the scale of this particular work, I am doubly grateful to have someone to talk it over with. So, be as critical as you want, and please feel free to suggest additions or changes.

I'm sorry it took so long, but this book went through a fair bit of existential angst! I knew from the start what I wanted the scientific content to be, so after our conversation, I jotted down a quick list of the times and places where people would first have collided with each of these ideas. On the train back from Geneva, I roughed out characters to fit in every slot. I knew I would have to do a lot of research before I could crawl into minds whose external reality was so different from my own, but I was quite looking forward to that part. By the time I arrived in Milan, I had begun to think this book wouldn't be nearly as hard as I had assumed. That illusion vanished instantly the minute I sat down to write.

Never having attempted fiction before, I thought I should learn a little more about the craft, but everything I read on the subject confused me further. I came to discover all the things that were wrong with my plan: I hadn't fleshed out the characters enough. Their personal arcs refused to curve; there was no drama, no source of conflict. One book suggested plotting the main events in the life of each character, to see how much movement and growth there was. I tried that and found nothing. No one went anywhere, did anything, they were all just there—thinking. Just when I was ready to give up, I saw it. The narrators weren't moving around or evolving, but physics was! With that shift of perspective, the plot points I had written down revealed a complex character arc: a fascinating journey, full of agonizing conflicts and moments of joy and fulfillment.

For a brief moment I struggled with point of view—that's another thing writing books bring up a lot! I considered letting Physics be the narrator, but that just didn't sit right with me. Physics might be the protagonist, but when have we ever seen or even heard from her directly? We know her only through interactions, the way tangents know curves; I wanted that to be reflected in the narration. From then on, the words began to fly, and a few months later, here I am.

One last word of explanation. I have been quite selective, almost idiosyncratic, in the subject matter I included. This was never meant to be a textbook, and whenever I attempted exhaustive explanations, the prose just sank under the weight; so I cut out everything that didn't fit in with the spirit of the writing. All I want my readers to know is the feeling of crazy ideas running rampant in a mind, changing its thoughts.

With that long preamble, here's some background on the chapters themselves. The first is set in the English countryside. In the year after Newton's death, the teenage son of a local squire aspires to comprehend the abstruse *Principia Mathematica*. It is the dawn of a new age. People begin to take their cues from observation instead of elaborate and obscure philosophy. Released from the grandiose constructions imposed on her for centuries, Nature is finally allowed to speak for herself. While many were part of this intellectual movement, Newton stood out, even in this distinguished crowd. His genius lay in his ability to translate observations from the vagueness of common language into the ultimate precision of mathematics. For the first time ever, conditions and relationships could be described exactly and then used to make quantitative predictions; the way that changed science is hard to overstate.

We have become blasé about classical mechanics, but when we consider it to be old and obvious, we do ourselves a disservice; like all great truths, these equations are not yet spent. People have fallen in love for millennia, but every time someone finds it, love is new again. The gems that lie in the depths of Newton's deceptively simple formulae gleam now just as they did centuries ago.

The second chapter is a letter written by a young Cambridge scholar to his artist sister. The Victorian society in which they lived was enamored with invention and innovation. Scientific lectures at the Royal Institution were social events, as men in pompous tailcoats and ladies with glamorous hats came to hear about the latest discoveries and watch demonstrations of this newfound knowledge.

I spent some time in London before I wrote this chapter. I visited the Royal Institution and peeped through a round glass window into Faraday's laboratory. As I wandered through the exhibits in the basement, I saw perfect orbs of hand-blown glass, tightly wrapped coils of wire, and faded labels on colored vials full of powder. Flasks, carafes, and bottles shimmered in their boxes. A dull metal shell sheathed a globe, like a pearl in an oyster. Gears and cogs of polished alloy were punctuated with dark wooden stands. I was spellbound. Whether you call them crude or elementary, these devices had a raw urgency and an undeniable organic beauty. When this equipment mediated our discussions with nature, the conversation was not hidden in silent, invisible, digital processes as it is today; it was manifested tangibly, through shakes and shivers, whirrs and thuds, sparks and flashes. I imagine the experience was akin to watching the planchette dart across a Ouija board as the summoned spirits spell out your Fate.

One of the greatest gifts of my research is a newfound acquaintance with Maxwell. The man had an amazing breadth of intellectual concerns, from larger philosophical issues to the way science should be conducted and communicated. On this last count in particular, I learned a lot from him. Maxwell wrote without pretension. He felt strongly that scientific histories should include descriptions of unsuccessful inquiries, and his own accounts are very much in that vein. Decades before the conversation about different learning styles, Maxwell advocated presenting knowledge in as many ways as possible, because "when the ideas, after entering through different gateways, effect a junction in the citadel of the mind, the position they occupy becomes impregnable."

With all this, and more, to choose from, what impressed me the most about Maxwell was his desire not just to teach science, but also to teach people how to think, through science. It is hard to conceive a more powerful use for the discipline. "My duty is to give you the requisite foundation and to allow your thoughts to arrange themselves freely," he told his students. "It is best that every man should be settled in his own mind, and not be led into other men's ways of thinking under the pretense of studying science." He talked about how a careful study of the laws of nature encourages "a habit of healthy and vigorous thinking which will enable us to recognize error in all the popular forms in which it appears and to seize and hold fast truth whether it be old or new."

Having read both Newton and Maxwell, I was struck by the contrast in their styles. Newton's sentences are concise; his emphasis is on mathematics, precision, and on demonstrating the correctness of assertions. The images he presents are tailored to perfection; all evidence of the process has been removed completely. Maxwell's language is lucid and fresh; his tone is authoritative but informal. In his writing, equations spring naturally from arguments. Some of these differences arise from the times and cultures to which each belonged, but others simply reflect their distinct personalities. Eliminating these characteristics is like reducing a photograph with deeply saturated color to black and white.

Do write back once you've had time to read these chapters. No rush, of course, but I'd really like to hear what you have to say.

All the best,

Leo

CHAPTER 1
Disturbing the Universe

[Classical Mechanics]

Do I dare disturb the universe? —T. S. ELIOT

NOVEMBER 1728
SHIRE, ENGLAND

I.

We need make haste no longer. Home is nigh. Safe in the knowledge that the stars will peek at us through the shutters near our hearth fire, we loosen the reins of hand and mind, let our tired horses amble along and set free our thoughts to wander.

Gently sloping hills in the distance are gilded by the declining sun. Fleecy clouds dapple the flaming sky, and the air is resplendent with harmony as birds sing their way back to their nests. Charmed as I am by the embellished prospect before me, my thoughts turn to the grove of trees on the wayside and the murmuring spring that lies deep in the darkened shadows beyond. I know that the sweet gurgling waters will not be visible from my current location, yet the contemplation of that hidden treasure only serves to deepen my ineffable felicity. It is so whenever one is in possession of a delicious secret, when one is privy to what lies otherwise hidden.

If such a commonplace secret can occasion delight, I wonder what raptures would strike the mind when a Secret of Nature is laid bare. How must Sir Isaac Newton have felt when he first beheld the adamantine gates of the Empyrean and the exalted abode of the gods lay in shimmering splendor before him? He had spent long hungry nights performing feverish calculations with unrelenting care, laboriously building each rung of the ladder which he then ascended to reach divine heights; but, I wonder, had all those lonely years of contemplating the imponderable sufficiently prepared him for the moment when he finally stood at the entrance to the sublime dominion where All is revealed?

Through his *Principia Mathematica*, Sir Isaac has bequeathed to the ages the ladder of Reason he so painstakingly constructed. In the first two volumes, he sets forth the mathematical principles upon which he builds his approach to natural philosophy; and in the third and final volume, entitled *The System of the World*, he uses them to lay bare the frame of the Universe in all its golden glory. The strength of these mathematical propositions is such that Sir Isaac is able to derive "the forces of gravity with which bodies tend to the sun and the several planets . . . the motions of the planets, the comets, the moon and the sea."

To gain these heights is my dearest dream, but in order to comprehend the wonders of the third Book, I know I must first master the principles established in the preceding two. Upon occasion, I weary of my slow and solemn task, and grow impatient for the time when I, too, will be able to handle these propositions with such finesse and grace that I may construct from them a model of the Heavens. But I have yet weathered only fifteen summers, and as Father says, precious tools and powerful weapons are not given into the hands of those who lack sufficient training and experience.

Sir Isaac knew the same must hold for forceful mathematical truths as well, so he erected a bar to the uninitiated: a wall that must be scaled in order to reach the treasure. The first two books of the *Principia* detail how this barrier may be surmounted, but since the task requires both care and diligence, Sir Isaac was assured that the sacred knowledge he had gleaned would be accessed only by those who could appreciate its strength, not by those who would subject it to "ridiculous comments or fool-ish disputes" simply because they were unable to "lay aside the prejudices to which they had been many years accustomed."

When I grow disheartened, I remind myself of the nobility of my goal and return to my task. As the Sun fixed on his majestic throne guides revolving worlds in harmonious orbs, the *Principia* guides my thoughts. Attracted first by one word then drawn to another, I wander to and fro over the text like a satellite agitated by the pull of surrounding planets, until, after multiple readings, I slowly become familiar with the orbit of thought prescribed in the sage lines. The conjoint influence of the Propositions steadies the axis of my whirling mind and keeps it from flying off into the deep spaces of uncertainty that surround its orbit.

Father has watched as I spent all day bent over my books. Often, I had no notion that night had fallen, until he came and opened the shutters of my closet. Understanding ebbed and flowed; as with the tides, there were periods of flux, when things became clearer and I felt my mind move in conjunction with another, higher Truth, and then there were periods of reflux, when no idea would draw my thoughts in and guide their course.

Father was gladdened to see me persevere; acquisitions not gotten with the rapidity of intuition are the more valuable, he said, as they are thoroughly made and firmly secured. He read to me Sir Francis Bacon's enjoinder to "proceed regularly and gradually from one axiom to another, so that the most general are not reached till the last: but then when you do come to them you find them to be not empty notions, but well defined, and such as nature would really recognize as her first principles, and such as lie at the heart and marrow of things." I knew Father was pleased with my efforts, but I did not know just how greatly, until this afternoon when I discovered the reason behind our excursion to town. I was truly surprised, and quite beside myself with joy, when Father told me that we were going to fetch a copy of Henry Pemberton's new book that he had sent for, from

London. Long had I wished for a gentle guide to steer me on my voyage through the sometimes harsh seas of the *Principia*, and by most accounts, it seemed that this was exactly the role *A View of Sir Isaac Newton's Philosophy* performed.

It is this precious parcel I now clench firmly in my hands. In his Introduction, Pemberton writes that he does not doubt Sir Isaac's wisdom in addressing the *Principia* only to mathematical readers, for in those early days, only those who might comprehend Newton's geometrical method of reasoning could be persuaded of the truth of his great discoveries; but now that Sir Isaac's doctrine "has been fully established by the unanimous approbation of all who are qualified to understand the same," Pemberton finds himself of the opinion that "such young gentlemen as have a turn for the mathematical sciences [should be encouraged] to pursue those studies the more cheerfully," and it is to facilitate their progress that he expounds Sir Newton's philosophy further, aiming to render it more easily understood.

The horses stumble on a stone in the path, the force of which interaction changes my previously uniform state of motion and breaks the reverie I was in. I turn to Father. He smiles; a smile manifests itself on my visage also. Who is to say if mine is a separate gesture in response to his, or if it is the same smile felt by us both? For, so say the Laws of Motion, when two bodies move towards each other, though the motions appear distinct, they arise from but one cause. It is a single action, arising from the conspiring natures of both, that draws them closer together.

My thoughts linger on Newton's Laws of Motion, those fixed foundations upon which rests the changeless order of the world. The very words themselves, heavy with power, invoke a reverence of the sort sorcerers in ages past might have felt upon reciting incantations; but this new magic is deeper than the old, for its efficacy depends not upon the positions of constellations in the firmament and its practice need not be confined to the spring and the fall of the leaf. Laws are stronger than superstitions because they do not rely upon elaborate rituals but rather are free of all such encumbrances. Reason's glorious spells, if such one may call them, can be cast as well on us as on each star that shines on us in the night, and on all the immense distances that separate us from them. These new Truths bring into the light what ancient rituals and superstitions had shrouded in the darkness of mystery. Desiring again to feel their heady power, I repeat the Laws under my breath.

Law I: Every body continues in its state of rest, or of uniform motion in a right* line, unless it is compelled to change that state by forces impressed upon it.

Law II: The change of motion is proportional to the motive force impressed; and is made in the direction of the right line in which that force is impressed.

Law III: To every action there is always opposed an equal reaction: or, the mutual actions of two bodies upon each other are always equal, and directed to contrary parts.

Father reaches over and shakes my shoulder to nudge me out of my reverie. We are home.

II.

I first learnt about the wonders of Natural Philosophy through little bits of dancing paper. My memory of that evening is so strong that I can almost feel the fire burn again in my limbs. I had only just gained my eighth year when fever confined me to bed for what seemed like an eternity measured out in slow, equally uneventful cups of lukewarm Time. Swallowing was painful and food tasteless, and the patient spoons Mother raised to me met with pursed lips. I had reconciled myself to a life of perpetual boredom and listlessness, when, drifting out of an uneasy sleep, I felt a strong hand smooth my brow and awoke to see not Mother, but Father by my side.

*The phrase "right line" was used to describe a straight line.

He bade me sit up, and with an air of great mystery, brought out of his pocket a small glass disk, a piece of cloth, and several little shreds of paper. These appeared quite unremarkable until Father rubbed the disk with the cloth and waved it over the paper bits. Without warning they sprang to life and leaped unto the glass; some moved perpendicularly, others traced oblique lines; some paused on the glass a moment, others sank down to rest; some leaped up again, others jumped up and then down, often turning around nimbly in mid-air as if caught in an unseen whirlwind. I was entranced by the spectacle and clapped my hands in glee.

When I had made him repeat the whole twice over, Father said I could keep the paper dancers for myself, if only I would finally take some sustenance. As I ate, he talked to me about electricity, that invisible power gained by bodies when they were excited, as by rubbing. Only trial will reveal whether a substance has this wondrous quality, he said. So we make the attempt, and then pass on the knowledge we have gained to others, by inscribing our transaction with Nature and recording our observations and reflections. So spellbound was I by this revelation, that I finished my bread and butter without being aware.

When Father stood up to leave, he rumpled my hair and said, "I think you would make a fine Natural Philosopher, John, if only you had the physical strength to go outdoors and observe the world as it turns. Every phenomenon carries within itself a Secret, which can be gained by anyone who asks Nature the correct Questions and is able to unravel her Answers. The whole world is just a giant riddle."

That was the beginning. Unravelling the mysteries that lay hidden around me was a noble cause, worth eating supper and building up my strength for. From then on, I became a committed Observer of the World, watching Nature's every little performance with the utmost zeal and interest, watchful for any clues She might let drop by mistake. In the evenings before supper, when his duties as landlord and farmer had been performed,

Father would call me to his library to ask what I had discovered that day. He listened with great solemnity and patience, and asked questions that steered my thinking. He pointed out to me the unspoken assumption upon which our entire system of thought rests: that what has happened once will, in similar circumstances, happen again. The events of Nature are not random. They follow a pattern. The future arises from the past as inevitably as if it were the product of a well-oiled machine. Once we are convinced of that truth, our quest to uncover the mechanism is inevitable.

Father and I talked about sundials and how the shadows of the linden trees in our garden painted the passage of time; we burst light open by holding up a prism to the Sun and letting the huddled-up colors disperse; we rolled marbles along the smooth floor, and up and down inclined planes made from books held open. Father told me to mind the similitude in things, and to discern the particulars wherein they disagree. He shared with me choice nuggets from books I was not then able to read on my own, but which have, in the years since, become familiar and beloved friends.

Bacon's vivid words made a deep impression on my childlike mind when Father read them out to me: "as an uneven mirror distorts the rays of objects according to its own figure and section, so the mind, when it receives impressions of objects through the senses, cannot be trusted to report them truly, but in forming its notions mixes up its own nature with the nature of things." At a young age, I learnt to recognize that our senses are fallible and our intellect prone to error. Natural Philosophers must always be aware of this bias, so that they may make allowances for, and perhaps even correct, their involuntary insidious actions.

Bacon reminds us that the goal is not to "guess and divine, but to discover and know," and if we do not propose to devise fabulous worlds of our own, but rather "to examine and dissect the nature of this very world itself," we must always return to the facts, "for it is in vain that you polish the mirror if there are no images to be reflected; and it is as necessary that the intellect should be supplied with fit matter to work upon, as with safeguards to guide its working." Sir Francis cautions us strongly, lest we "give out a dream of our own imagination for a pattern of the world."

The only way to avoid this, Father explained, is to assume no greater knowledge than that which one finds himself rightly and indisputably possessed of; to indulge in no conjectures concerning the laws and patterns of nature, but instead seek with all diligence, by gradual inquiries, the true laws which regulate the constitution of things. Such has not always been the case, Father explained. It was the custom of old for philosophers to frame conjectures, and declare them to be true should they show even an imperfect agreement with Nature. "Perhaps it was the grandeur of the quest that kept them from thinking they could ever have a true and complete answer," Father sympathized. But in the ages since, Natural Philosophers have begun to guard against the great absurdity of proceeding upon conjectures. They are loath to make hasty transitions from first observations to general axioms, and instead advance "with the utmost caution, and by very slow degrees," as Pemberton wrote. Only when large stores of evidence have been amassed do they attempt to synthesize knowledge and extract from it a common foundation.

Once assured that a given property holds in every situation open to exploration, Natural Philosophers proceed to assume it is true in similar situations universally. This process of induction is extremely important, for without it, no progress could be made. We have no way of finding the properties of such bodies as are out of our reach, and so, we take what appears to be true in every known case and elevate it to a universal, though provisional, truth. For, Father reminded me, as scientists we are to frame no hypotheses nor receive them into our philosophy otherwise than as questions. Even that which we take as truth may be disputed, should further discoveries expose its contradictions or limitations.

Pemberton explains it thus: proofs in mathematics can be conclusive, for in that science, we deal with the ideas of our own minds, "so that as the mind can have a full and adequate knowledge of its own ideas, the reasoning in geometry can be rendered perfect. But in natural knowledge the subject of our contemplation is without us, and not so completely to be known."

Father told me about the Royal Society in London where a group of men meet weekly to witness experiments, and discuss and debate issues in natural philosophy, with a view to the advancement of valuable learning. In service of this noble cause, they collect and receive information about discoveries new and old from every quarter of the Earth. They believe the grand design of the world too vast to be uncovered by any one private Writer thinking alone in the darkness, and thus disseminate their collated information in a paper called the *Philosophical Transactions*. What impressed my child-like mind with the strongest force was their motto, *"Nullius in verba,"* the bold declaration to accept nothing on mere authority, but rather to verify all statements by experiment. No man is accorded more deference than another. All claims, no matter who makes them, are subject to test. We are as equal in the eyes of the Royal Society, said Father, as we are in the eyes of God.

Imbued with the same spirit, I carried out my various experiments, such as they were. So taken was I by Galileo's assertion that all sorts of heavy bodies descend to the Earth from equal heights in equal times, that to Mother's intense and often comic dismay, I dropped various objects of all sizes and shapes from the garret to the ground below. I looked hopefully for deviations from the law, but objects that were dropped simultaneously always kept company through flight and hit the ground together. At length, Mother prevailed upon me to accept this as fact and to investigate other laws.

I peered long and hard into phenomena. Some were compelled to give up their secrets under the strength of my stare; others, I modeled and recreated, in an effort to understand how they came about. I built clocks and made kites, and when I first saw a windmill in a neighboring town, I came home and constructed a small replica, to learn how it operated. Bacon said that Nature gives up her secrets not only when she is "left to her own course and does her work her own way," but also when she is "under constraint and vexed," when "she is forced out of her natural state, and squeezed and moulded." Keeping his words in mind, I changed my designs to see if I could make my models more efficient. Of the kites, I made various sizes and shapes, carefully noting which flew better than others. When these constructions were done, I would lie for hours among the asphodel and moonwort, staring at the sky, trying to puzzle out the reason why things were so. Soon, I realized that I would need to preserve my observations or risk forgetting them. Accordingly I stitched a few sheets of paper together and bound them in vellum to create a notebook. Dipping my quill in the sea-blue ink of dried, bruised privet berries, I poured out my facts, thoughts, and conundrums. On line upon line, I recorded what I had learnt, what worked, what puzzled me. It was as if I had compiled my own little miscellany, a collection of curiosities tied together by my personal tastes and preferences.

One evening, when the pages were nearly full, I showed them to Father. He quite liked my effort and said it proved I was ready for a real notebook, one I could confide in as in a friend, one that would serve as a perpetual record, should memory confound or betray me later, and most importantly, one that had enough empty pages to grow with me. "Pen all your findings and reflections, John," he said. "Do as Bacon did, and 'proscribe and brand by name' all falsehoods you encounter, 'that the sciences may no more be troubled with them.' When you make an experiment, keep a strict and clear account, so that others may see 'how each point was made out, may see whether there be any error connected with it, and may arouse themselves to devise proofs more trustworthy and exquisite, if such can be found.'" As Father went on to tell me, the late President of the Royal Society, Robert Hooke, said in his *Micrographia* that, even when a man's remembrance is imperfect, it regulates his actions; thus is a physician accounted the more able because he has had long experience and practice. What then of the man "that has not only a perfect register of his own experience, but is grown old with the experience of many hundreds of years, and many thousands of men"? "When you record your results for posterity," Father counseled, "remember to design them for public use rather than for ostentation, and do not let vanity persuade you to omit your more unfortunate experiences."

So saying, he brought down a handsome volume from his shelf. It was fully a thousand sheets, bound in beautiful brown leather. Taking his quill, Father wrote on the first page in his fine flowing hand, "This book is the property of John ——." Happiness spread through me like the warmth of a fire on a cold winter's evening. Father looked up at me, and I knew he could see the glow within. Before he gave me the book, he inscribed on the flyleaf, *"Nullius in Verba."*

My happiness was complete.

III.

As my investigations into natural philosophy slowly became more sophisticated, Father began to introduce me to some of the grand new ideas in Mathematics. He told me the tale of fluents and fluxions to illustrate how a slight adjustment in thinking can sometimes open up a whole new field of view. We both knew that, at the time, the details were beyond my grasp, but I enjoyed the story nonetheless and the underlying thoughts stayed with me.

I had already noticed that there are few uninterrupted straight lines in Nature. Most objects are curved in their material extension, and in the paths they travel. Cannonballs trace parabolas in the air before they fall to the ground, and planets execute their celestial motions in ellipses. But these omnipresent curves presented a deep challenge to the mathematician: how could a curve be represented as a single, smooth object? The prevailing model, prior to Newton's time, was to approximate a curve by a series of small straight lines of changing orientation, placed end to end. The smaller the size of the lines, the more closely did this construction resemble the curve. But, regardless of its success, this treatment fragmented a curve, breaking a unity into a series of jagged lines and edges.

Newton approached this problem from an angle that deviated from the common in an apparently slight but ultimately crucial manner. Rather than considering a curve to be pieced together by a collection of small lines, he perceived it to be the shape traced out by a single small line which constantly changed direction. To him, a curve was not a static object, but an evolution.

It was already well known that when quantities evolve in time, a determination of the final outcome requires knowledge of both the amount of change and the velocity with which this change takes place. The same logic should apply to a quantity that evolves across space, Newton said. *All* change describes a flow, he explained, and there is always a velocity associated with flow. The quantity undergoing the flow, Newton termed the fluent; the rate of flow, he called the fluxion.

Applying these ideas to a curve, Newton started by shrinking the very small line down as far as possible, until it became a point. Taking this point as fluent, and being in possession of the corresponding fluxion dictating its evolution, Newton could construct the curve that represented the smooth flow of this one single point across space. On this foundational insight, he erected the entire frame of the calculus. That there is a science to study flux is a good and useful thing, for truthfully has it been said that nothing endures but change.

I had not yet attained my tenth year when Mother was wrested from us. In the bleakness following her death, sorrow heaved and swelled like a rolling sea. Grief took root within Father's soul, and I began to fear I would lose him also. The fire within them extinguished, shadows appeared on his eyes, making it clear they were suns no more, and could shine now only in the borrowed light of a star.

My Uncle William, the Rector, came to commend Mother to her Maker. She had been his favorite sister and he was bereft as well, but in his Faith lay his strength, and ours. Uncle William and Father had chummed together during their days at Cambridge and remained close ever since. In matter of fact, that was how Father met Mother.

After the funeral, when everyone had left, I saw Uncle William and Father sitting at the table, talking. Father held up a small likeness of Mother, made when they were newly married. The painting had all the beauties and graces a skilled artist could bestow, but Father would insist it did not do her justice and that one day he would have a more befitting portrait made. That awful night, as he clenched the picture in his hands, I thought I heard him sobbing. I did not know what else to do, so I softly ran up the stairs, climbed into my bed, and embraced the darkness until sleep finally took mercy on me.

All through the next days, visitors poured in. Mother had been popular among the tenant farmers, and they all came to offer their condolences. I escaped into the garden to avoid the inevitable looks of pity. When Uncle William found me, I was in her bower, the one she loved especially, looking up at the clouds as they floated by. Without a word, he lay down beside and watched the sky with me. After a while, his eyes still on the clouds, he spoke:

"Has your Father ever mentioned Nicolas Saunderson?" The name was vaguely familiar. Saunderson had been a Professor at Cambridge when Father and Uncle William were students there. "It was from him that your Father and I first learnt of Newton's theories," Uncle William said. "His knowledge and popularity were extraordinary, but what made them truly astonishing was that he was blind. Yet, because he persevered in his endeavors and did not succumb to the pain of his position, he was able to behold, with his mind, the majesty of the great universe. The hidden secrets of Nature were seen plainly by his darkened eyes, and ultimately he succeeded to the Lucasian Chair formerly held by Newton.

"There is great power and peace in the contemplation of the natural order," Uncle William continued. "It influences the temper of your mind. You come to see that this most beautiful universe could only proceed from the counsel and dominion of an intelligent and powerful Being whose glory the Heavens declare. The world changes constantly, yet in the midst of this flux, there is a quiescent center—a system of the world that transcends Time. Anchor yourself to that and you will better weather the storms of life. Your loss is truly great, but should you let them, the beautiful harmonies of nature will help fill the aching void in your heart."

Silence crept in to fill the space between us, and we lay for a while, thus cushioned. Eventually, Uncle William stood up to leave. As he brushed the grass from his coat, he said, "Your Father loved your Mother so. A powerful bond held them together, and now that it has broken, I fear he will go down this straight path of pain, unless some force intervenes to steer him otherwise. You are his Sun, John, the brightest source of light in his sky. Your love must be the centripetal force that reels him in, the power of gravity that tends to him, keeping him stable in his course, speeding through his days with regularity and purpose."

I nodded. Uncle William chucked my chin and walked back indoors.

I crept up to my closet unseen and sat down to think. That it is the inherent nature of material bodies to oppose a change in their state of motion, I knew. This was Newton's First Law, and Father and I had discussed it often. I understood that, left to themselves, all bodies persist as they were, maintaining their state of motion, whether at rest or moving with constant velocity. The quantity of matter in a body is but a measure of this *vis inertiae* or quality of inactivity. Any change in the state of motion of a body, whether it be the change from rest to motion or motion to rest, or a change in the direction of motion, or even the change from one definite velocity to another, must always signal the presence of a force, for nowise would a body perform this action of its own accord.

I also knew, from Newton's Second Law, that an impressed force would alter the state of motion of a body by an amount proportional to the power exerted; and that the same force would cause a lesser change in the motion of a more massive body and a greater change in the motion of a less massive body. The change wrought is without regard to the motion the body already possesses. The same power will always produce the same degree of change in the motion of a given body; regardless of whether the body move from rest into motion, or from one state of motion into another, the difference between the initial and final states will remain the same.

This much I was sure of, but the mechanism of gravity was unknown to me. And how should I wield a force if I knew nothing of it? I drew out my beloved notebook. The filled pages called out of things that were certain and unchangeable, and in that there was comfort. Empty pages beckoned with promise, whispering of the secrets that might be inscribed there in years to come. Somehow, I felt much better.

Turning to a blank page, I wrote in large letters, "Centripetal Forces and Gravity."

IV.

In the words of Sir Isaac, "A centripetal force is that by which bodies are drawn or impelled, or any way tend, towards a point as to a centre. Of this sort is gravity, by which bodies tend to the centre of the Earth; magnetism, by which iron tends to the loadstone; and that force, whatever it is, by which the planets are continually drawn aside from the rectilinear motions, which otherwise they would pursue, and made to revolve in curvilinear orbits."

The Earth has orbited the Sun six times since I copied out those words in my childish hand, but my mind has spun around its axis innumerable times and my understanding has been much augmented in the process.

In the dark days after Mother passed, I took Uncle William's counsel and made a deliberate effort to spend more time with Father. I began to be less shy of seeking him out, for I started to see that our conversations were a necessary and purposeful diversion for him as well. Most evenings, we talked of our mutual joy in Natural Philosophy, and as the seasons passed, both my understanding and my bond with Father grew stronger. I supplemented these discussions with solitary hours spent in laborious thought, and chronicled my philosophical journey in my faithful companion notebook. As I look back over the pages now, I see my struggle made visible in the carefully written lines as well as the scribbled notes in the margins.

Gravity has not been easy to comprehend. In the beginning, this much only was clear to me, that gravity made objects fall. Being a force, gravity must necessarily change the state of motion of an object in its thrall, so it followed that the velocity of a falling object must change constantly. I knew, from Galileo's assertions, and my own various experiments to this end, that the Earth's gravity caused any two objects dropped from the same height to reach the ground at the same time, regardless of their mass and extent. Being that they cover equal spaces in equal times, the objects must always travel at the same velocity. Because these velocities change, they must change in an identical manner. This was the peculiar attribute of gravity: that it changed the state of motion of every object in the same fixed manner, paying no regard to mass or extent. Clearly, then, the strength of this force could not be fixed.

Even when I accepted its variable nature, there was much about gravity that troubled me. What, for instance, was the similarity between an object falling to the ground and one suspended in orbit? How could these various motions be attributed to the same cause? Doubt gnawed at me with its little mouse teeth, whispering that the descent of an apple to the Earth was in no way similar to the graceful arcs traced by the Moon, but I put that out of my mind and concentrated on the facts. The Moon orbits the Earth, tracing a curve of known radius and extent, in a fixed period of time. As this motion is not linear, we know from Newton's Laws that a force must be present. Only under an incessant influence would the Moon be deflected, continually, from its tangential inclinations and held in orbit. Just as a stone in a sling is straining to fly off, writes Sir Isaac, so is true for all bodies in orbit. "They all endeavor to recede from the centers of their orbits; and were it not for . . . a contrary force which restrains them to, and detains them in their orbits, . . . would fly off in right lines with a uniform motion."

What kind of a force would compel circular motion? In diagrams furiously and repeatedly sketched, I tried to convince myself that a body in a circular orbit may be construed as being in perpetual fall under thrall of a force directed radially inwards to the center of the orbit, but understanding did not truly come until I started playing with a wire, bending and turning it with my own hands. I recorded my experiment in my notes thus:

"If I lay out a wire in a straight horizontal line and leave it alone, the wire remains unchanged. But if I hold it, a little distance from the end, and push it down, the wire becomes angular. If I exert this downward force once, the wire is bent only at a single point; from that point forward, it continues as straight as it was before, only now it is inclined towards a different direction. I can then turn on an angle, shift my vantage point such that the straight extension looks to me horizontal, and repeat my previous action. If I push downward again, at a little distance from the first bent point, employing the same force as earlier, I create another angle, and from thence the wire extends straight ahead till the end, though its direction has yet again been modified. Once more, I turn, until this straight segment appears horizontal; if I repeat this process until the ends of the wire meet, I have moulded my wire into a polygon. The smaller the sides of the polygon, and the slighter the angles, the closer this shape is to a circle."

Thus was I finally able to see that when a force acts on an object continually, pulling it down towards itself, the object moves in a circular orbit. Only when I learned to decompose an orbit into a perpetuity of small falls in constantly changing directions did ghosts of similarities between centripetal forces and gravity finally float in front of my eyes. Vaguely, I began to see that if a centripetal force be that by which bodies are impelled, or otherwise tend, towards a point as a center, then gravity, by which bodies tend to the center of the Earth, was a centripetal force that could keep the Moon pinned in orbit.

I searched for these faint forms in all the places I was told to look, and even yet it took me several days to spy them. Newton's genius was such that he sensed their presence in the dark, and lured the phantasms into the light of day. Newton believed that order underlies apparent complexity and that in cause, though not always in effect, Nature is simple and consonant to herself. In this spirit, he laid out a set of rules, establishing a canon for Natural Philosophy much as there is for the Church. The rules were simple: The primary causes of all things should be derived from the simplest principles possible, for causes ought always to be simpler than their effects. To like effects, the same causes are to be ascribed; else nothing can be affirmed as a general truth. More causes are not be to received in Philosophy than are sufficient to explain appearances, for Nature "is pleased with simplicity and affects not the pomp of superfluous causes."

And so, when Newton was faced with two apparent forces, both of which exerted a pull toward the Earth, his belief in Nature's economy caused him to wonder if there was a deeper reason behind the coincidence; if these two forces we called different, were somehow the same.

When I first heard this, I was plagued with confusion. If both are subject to gravity, why then does an apple fall to the ground while the Moon traces circles in the sky? Why does the apple not fly into orbit, while the Moon falls to the ground? Undeterred by these superficialities, and ever methodical in his progress, Newton started by establishing a link between the vertical fall of an apple and the circular orbit of a satellite.

We know from common experience that an object dropped vertically continues its descent in that direction, but an object propelled forward (or indeed at any angle other than the vertical) meets the ground in an arc. Such is the case for cannonballs and all other projectiles; they trace parabolas in the air, as they hasten to answer gravity's siren call. Could they ignore gravity, these leaden balls would shoot off in straight lines, and maintain the velocity imparted to them by the gunpowder. The larger this initial velocity, the closer a ball keeps to its rectilinear course and the farther it goes before its eventual fall. Indeed, we see that cannonballs fired with greater force travel a longer distance along a more generous curve, before the surface of the Earth intervenes and cuts short their flight.

"If a leaden ball, projected from the top of a mountain by the force of gunpowder, with a given velocity, and in a direction parallel to the horizon, is carried in a curved line to the distance of two miles before it falls to the ground; the same with a double velocity would fly twice as far," Sir Isaac wrote. And if the velocity were increased tenfold, the distance would increase in like proportion; in fact, "by increasing the velocity, we may at pleasure increase the distance to which [the ball] might be projected, and diminish the curvature of the line which it might describe." Then, by induction, we may conclude that if the ball were imparted enough velocity, it might move in a curve so large as to circumscribe our planet. The radial pull of Earth's gravity would keep such a ball in perpetual orbit, as if it were a satellite.

If this be true for any satellite, Newton wondered, could it not also be the mechanism behind lunar motion? Might a force emanating from humble Earth hold the silver-tressed Moon in orbit?* It was known that gravity reached the summits of the highest mountain tops, for things fell down even from there, but exactly how far did it extend? If gravity did stretch out into the Heavens, would not its power be strained, and grow ever weaker?

Others before him had suspected the existence of such a diminishing force, but none had been able to do more than conjecture. Unlike his predecessors, Newton was not satisfied with such vagaries. He employed exactitude in thought and calculation and sought a confirmation of his conclusions by Nature who, alone, was his judge. Rather than writing down elaborate ideas cloaked in entangled words, Newton wove an intricate mathematical web that enabled him to ensnare the force of gravity and study its characteristics. He made concrete predictions that could be measured and verified.

Using the precise, and indubitable, tools of Mathematics, Newton calculated the velocity that must be imparted to a cannonball if it is to circle the Earth. Once this correspondence was framed in terms of geometry, the relationship between a satellite's velocity and the size of its orbit became manifest. Lo and behold, the motion of the Moon fit this description like a glove. A satellite moving with the same velocity must necessarily follow in the footsteps of the Moon. Upon the strength of mathematical propositions Sir Isaac proved, unequivocally, a fact resembling poetry: that the Earth swings the Moon upon the invisible rope of gravity, like a child whirling a stone in a sling, around and around.

Having obtained success in this one instance, Sir Isaac forged ahead. The Heavens are perpetually in motion, he said, and if

---

*Newton went one step further, conjecturing that if we imparted enough velocity to an object, it "might never fall to the earth, but go forwards into the celestial spaces and proceed in its motion, in infinitum."

the Moon is retained in her orbit by the Earth's gravity, should not the planets be urged in their similar motions around the Sun by a like power?

To investigate these ideas, Newton leaned upon the work of the learned men who had come before him, a chain of harmonious minds linked together by reason. The labor of Copernicus, Galileo, Brahe, and Kepler saw its full fruition when Newton used their painstaking computations, and their detailed descriptions of the Moon's orbit, to calculate the force needed to keep it in such motion. A fine balance had to be maintained. As Newton noted, "If this force was too small, it would not sufficiently turn the moon out of a rectilinear course: if it was too great, it would turn it too much, and draw down the moon from its orbit towards the earth. It is necessary that the force be of a just quantity and it belongs to the mathematicians to find the force, that may serve exactly to retain a body in a given orbit, with a given velocity."

In his explorations of planetary motion, Johannes Kepler had discovered that the times taken by different planets to complete their orbits were in direct relationship to their several distances from the Sun. Using indisputable geometrical principles, Sir Isaac demonstrated that this particular relationship could be caused only if the centripetal force on the planets was related to their distance from the Sun, in the same proportion as that which exists between the side of a square and the inverse of its area. If the Moon's orbit was ascribed to the Earth's centripetal force, this power should also decrease in a similar manner.

Since he knew the tug it exerts upon the Moon, and the manner in which it diminishes, Newton was able to calculate the strength of this force at the surface of the Earth. This turned out to be none other than the familiar pull of gravity, making it certain that the force underlying lunar motion *was* gravity. Had the forces been distinct, the influence on a falling object would increase twofold, causing it to descend twice as swiftly as it is seen to do. This calculation further strengthened the argument that the mechanism which caused the Moon to orbit the Earth was the same as that which dictated the Earth's orbit around the Sun.

From these facts, Newton induced the principle of Universal Gravitation: namely, that a gravitational attraction exists between any two massive objects, such that they are "heavy towards each other." Having stated the fact, Newton set out to demonstrate its power. He used gravity to explain all manner of diverse phenomena. Astronomers had long been puzzled by the precession of the Earth whereby, every seventy-two years, the orientation of the Earth's orbit changes by one degree. Newton showed that this previously inexplicable occurrence was an inevitable outcome, the result of the confluence of the gravitational pulls of the Sun and the Moon upon Earth.

If, as the Law of Universal Gravitation claimed, each object attracts every other object, with just regard to its mass and the distance between them, then the orbit of each planet should depend upon the combined motion of all the planets and their actions upon each other. Newton showed that this implication, too, was borne out. The disturbance that appears in the orbits of Jupiter and Saturn when they near their conjunction was but the observable effect of their mutual gravitation. The Law dictates that bodies not glide insensibly by each other in vast space, but in deference to one another, modify their courses and modulate their conduct.

As a lens focuses the different colors of light, combining them again into the white beam from which they were refracted, Newton collated various phenomena that, when viewed through the prism of our experience, appear separate, and made them one again. In the brilliant ray of Universal Gravitation, many diverse motions are subsumed.

The tides in the sea were explained as being due to the Moon's pull on the waters of the Earth. Comets in the sky were exposed as being objects bound to the Sun through long series of revolving time. Their blazing tails of fire no longer spread tremendous with imagined ills; no more were they considered "prodigies of fear" and "portents of broached mischief to the unborn times"; their flaming appearances portended naught but their tedious passage in orbits decreed as surely as ours around the Sun. They were not harbingers of the fates, and as such, neither to be feared nor consulted in human affairs. We are the ones with the power to change our destinies; it is the heavenly bodies that are in fetters.

V.

The *Principia Mathematica* is the keystone of Natural Philosophy. It connects ideas long thought to belong to entirely different edifices and brings them together to create an Arch so glorious and complete, it could only have been dreamt of by a Divine and benevolent Creator—so I mused as I read the Preface again last night. As I marveled at the strength of the *Principia*, at the truth and precision of its every word, Father walked in. My studies continue to be a bond between us, one that grows stronger with each passing year. I believe that, to some degree, I have fulfilled my uncle's charge to act as a centripetal force and keep Father steady in his course. Finding me lost in thought, Father asked what I was contemplating, and I told him.

"Sometimes I wonder," he said, "if you, who have grown up in this new scientific tradition of Bacon and Hooke and the Royal Society, can fully appreciate the genius it took to compose the *Principia*. It is only when you compare the nature of this tome to the style and argumentation that dominated Natural Philosophy in the decades hitherto, that you truly appreciate its power." Walking over to his shelves, Father picked up an old volume entitled *The Memento Mori of Sir ——— Esq.* He dusted off the covers, leafed through the pages, and started reading:

"'Above is the right essence, below is only the Type and Image. Above are the right Principles and Elements; the below are only a shadow; the Terrestrial bodies are meer Ashes, but the Celestial are a noble salt of life.'"

And this: "'Always out of the upper things, things beneath were gotten, and the upper is always before that which is below, even as the Spirit is first before the Soul and Body.'"

"I will not read you the whole of this archaic pattern of thought," Father said, "but I want you to be aware of the mystical approach that prevailed when Newton was young. There was a hierarchy of importance, and the Eternal Heavens were held to be far superior to and vastly distinct from mortal Earth. The phenomena Newton studied were not new, but his perspective was. Galileo had made huge strides in understanding earthly mechanics, and Kepler had described celestial mechanics quite successfully, but these domains were thought unquestionably distinct, and no one had dared consider synthesizing them.

"Through his Law of Universal Gravitation, Newton made a claim that bordered on unacceptable for some: that the laws governing the star-glowing firmament were the same as those that caused the ebb and flow of our seas. The affairs of this terrestrial globe were considered together with those of the innermost Heavens; the ceaseless reciprocal influence of gravity linked body to body, the minutest to the stupendous, each to each, each to all and all to each. In a remarkably profound insight, he declared all objects in the universe equal under this influence."

"As the Royal Society declared equality among the opinions and observations of men," I said. Father smiled. "Yes, it was in much the same spirit. The terrestrial are not shadows of the Celestial, any more than the opinions of men of lowly station are ashes as compared to the observations of noblemen. The worth of an idea lies in its own intrinsic value, and the manner in which it withstands tests and demonstrations. Nature treats us all, King and peasant alike, according to the same rules. She distinguishes not between the Heavens and the Earth.

"The Law of Universal Gravitation," Father continued, "was so daring, and of such vast scope, that many rebelled, finding Newton's ideas too revolutionary and his methods too difficult. They criticized him for re-introducing the occult into philosophy. What was this mysterious action at a distance, they asked, if not an occult phenomenon? They pointed to Newton's own admission:

"'I have explained the phenomena of the heavens and sea by the force of gravity, but I have not yet assigned a cause to gravity. . . . I have not as yet been able to decide . . . the reasons for these properties of gravity, and I do not feign hypotheses. For whatever is not decided by the phenomena must be called a hypothesis, and hypotheses, whether metaphysical or physical, or based on occult qualities, or mechanical, have no place in experimental philosophy. . . . It is enough that gravity really exists and acts according to the laws that we have set forth and is sufficient to explain all the motions of the heavenly bodies and of our sea.'

"But these shortsighted critics mistake the *Principia*," Father said. "What was called occult in ages past was that which lay beyond our realm of comprehension, rendering it vain to seek a further cause. The unknown in Sir Isaac Newton's philosophy is nowhere claimed to be unattainable. In fact, just as much has been revealed to us that lay hidden to our ancestors, it is hoped that further experiments and patient enquiries into Nature will reveal the causes and mechanism of gravity to later generations.

"Beyond doubt, it would be more satisfying to trace all effects to their first original causes, but how are we ever to come to this knowledge otherwise than by storing up all the intermediate causes we can discover? By sufficiently having proved any one cause, a scientist lays a safe foundation for others to work upon, and facilitates their endeavors in the search for yet more distant causes. Until that comes to pass, even intermediate knowledge may be put to many useful purposes. Long before it was known that the ascent of water in pumps is due to the pressure of the air, the very fact of its occurrence was enough to construct devices that aided daily life. Even when we were unacquainted with the spring in Nature whence this power was derived, we were able to estimate its effects.

"It is the duty of every Natural Philosopher to find out what he can. Each step leads us nearer to the original cause. To neglect what you can do because you can advance no further is plainly ridiculous. Sir Isaac drew on a deep courage when he confessed the imperfections and incompleteness of his epic work and outlined what he did not yet know. He made neither apologies nor excuses, but simply laid the matter plain. Newton fully realized that gravity, as he saw it, was a force that acted across large distances, mediated apparently by nothing. It was not his failing but his greatness that he admitted this, and tasked future scholars with finding the explanations. He wrote, 'To explain nature is all too difficult a task for any one man or even for any one age. 'Tis much better to do a little with certainty and leave the rest for others that come after you.'"

Father grew thoughtful. He walked to the fire, stroking its embers, and paused there a long moment before he spoke again. "When I was young," he said, "my father, your grandfather, became much absorbed in chymistry. He even built a small room at the end of the garden, where he spent many an hour engaged in experiments. It was a fantastical place, and appeared even more so to my child's eye. Well I remember the brick furnace, the charcoal fires, the distillation apparatus, the moldy books, the jars of minerals, and the packets of nameless exotic substances. Often, at dinner, he would regale us with discourses on the beauties of nature as reflected in the chymicals in his tubes and crucibles.

"At length," Father continued, "word spread around the countryside that a gentleman who had his own laboratory was engaged in chymical experiments, and the news brought an adept to our door. He attempted to convince us that he was a man of great skill and that his experience in the operations of the Spagyrists had led him to discover the philosopher's stone, which could turn lead into the best and finest gold. Your grandfather was a wise man, and not tempted by avarice, so he turned the adept away. That evening, as he told us children about his day, he reminded us that the analyzing of bodies by fire was a praiseworthy task. The discoveries he made filled his mind with the greatest veneration for the wonderful Author of Nature. He wished for us to experience these same joys, but hoped we would resist the siren calls of Alchemy, for that sad romance ends only in empty pockets and lost hours.

"I have borne his words in my mind," Father said. "They have steered me true and served me well. So I bequeath them to you now: There are those who hold that Nature can be laid open only by the force of fire. Perhaps they are correct, but they mistake the nature of this flame. It is not apparent fire by which we must resolve Nature, but the divine fire of the mind, which will make all volatile opinions vanish into smoke, leaving only the affirmative form, solid, true, and well-defined."

Father's eyes were gleaming now. "Should you want to indulge in Alchemy," he said, "sift out the dross and transmute centuries of diverse observations into shimmering principles, for those are the only true nuggets of gold, and the only immortal elixir is the enduring knowledge you leave behind for others."

CHAPTER 2
As Though They Were Fairy Tales

[Classical Electromagnetism]

*A scientist in his laboratory is . . . [like] a child
confronting natural phenomena that impress him as
though they were fairy tales.* —MARIE CURIE

MICHAELMAS, 1893
CAMBRIDGE, ENGLAND

My very Dearest Lizzy,

Please believe me when I say I have meant to write to you these
many days, and I most sincerely beg your pardon for not stir-
ring myself to it earlier. All this week I have been thinking of
you, but by the time I return to my rooms each night, I am so
exhausted from my studies that the very thought of putting pen
to paper fatigues me and I resort to having conversations with
you in my head as I drift off to sleep. Tonight, however, I came
back to find Mother's letter waiting and that settled the matter.

Mother is quite in raptures over your betrothal to Julian and
thinks you two will suit perfectly. She tells me that Julian has
excellent prospects and that Father has sorted out the details
of the marriage settlements already. Now, while all those matters are important, they are best dealt with by Parents. What lies
at my door, as your brother, is to find out more about the man
himself. You will be pleased to know that I have discharged that
duty with the utmost discretion. Since he is a few years older
than us, none of my set knows him intimately, but in the judi-
cious inquiries we have made this evening, we have heard not a

single bad word. It would appear that he keeps an excellent stable, is a generous friend and an accomplished cricketer who led his team to victory at Lord's a couple of years ago! In short, other than the fact that he is an Oxford man, he appears to be a capital fellow, and so I wish you very happy.

While we are on this subject, one more thing. I know you love everything about London, from the glamour of the Ritz to the flower girls at Covent Garden, but if you ever find yourself lonesome or homesick, you have only to tell me and I can be there instantly. Among my varied acquaintance, I number several bloods who have mastered the art of driving to London surreptitiously in their tandems and sneaking back in to college after midnight curfew. Knowing that the thing can be done without getting caught, I shall be happy to oblige you with my company whenever you desire. Of course I would also be willing to make this gesture for Julian, should he be inclined to deepen our acquaintance over lunch at the Club, or in the event that he requires a companion for the Friday Night Discourses at the Royal Institution (Ned tells me Julian is a subscriber). And if this be the case, I solemnly promise, Lizzy, that I shall not be the scapegrace I may hitherto have been on occasion but, on the contrary, such an elegantly behaved young gentleman that I will do you proud. The more I think about it, the more I am convinced that Julian and I shall get on famously, for I always have wanted an older brother.

Not that I didn't enjoy having an older sister. In fact, since I received Mother's letter, I find myself quite nostalgic for the almost perfect times we spent together. If it is true that we are forever marked by our childhood, then you and I have been rather lucky. In the tide of memories that crashes upon the shores of my mind, there is no unhappy wave. I remember jokes and banter, fairs and dances and visits with neighbors; I see all four of us together at supper, talking about our day; I recall stimulating conversations by the fire in the library, with us two at Father's feet and Mother sitting close by; I think of you and me riding our ponies, of our rambles in the gardens, our skating expeditions; I remember playing with marbles in the portrait gallery, under the frozen gazes of our ancestors, and spinning tops on the family silver; I think of the plays and entertainments we put up, and of our visits to London when Mother would forgo shopping at Burlington Arcade to take us to comic operas, balloon ascents, fireworks displays, and museums with Father.

As I relive it now, I realize how large a debt we owe our parents. They awoke such a love for Nature within us, Father with his deep curiosity and his insatiable interest in every conceivable phenomenon, and Mother with her gentle reverence for it all. I will never forget Father quoting Michael Faraday—something about how there is no tranquility in the Cosmos, that the sea is at war with the earth, and the air lends itself to the strife. I remember so clearly how he told us to pay attention to everything, because the little processes we tend to pass over are in fact essential and powerful and absolutely full of wonder. And Mother, with her serene smile, would quote from some poet or other. I remember in particular one poem by Gerard Manley Hopkins, in which he said all things are charged with love, "and if we know how to touch them, give off sparks and take fire, yield drops and flow, ring and tell of Him." Obviously, Father's discourse struck a chord with me far more than Mother's, but I am beginning to think now that perhaps their different expressions are both equally important to the Truth.

I know that is not a tone I have ever taken before, and I am sure you are wondering if an excess of sentimentality has addled my brain, but I assure you that is not it. These thoughts have been in my head for a while and were the particular reason I was meaning to write to you long before Mother's missive came and made procrastination impossible.

Why is that? you ask. Well, think back to when we were children out for a walk in the gardens outside our house. Every now and then, one of us would spy a treasure, perhaps a speckled blue egg which carried within itself the promise of a sparrow, or jewel-toned leaves, in whose crisscrossed veins we would read maps to buried pirate's chests. Whoever first spied these riches would rush excitedly to the other, impatient to multiply the joy by sharing it. Some of my fondest memories are of you running across the grass, crying, "Look, Charles, look!" This letter was meant to be my grown-up version of that cry.

I remember how, after we got over the first happy shock of discovery, I would turn instinctively to books or pepper Mother and Father with questions about these mysteries; I had a burning need to know why things were so, while you would keep the objects in your hands, turning them over silently, until they turned into your wonderful paintings. I hope you intend to continue doing that, even when you are a fine married lady with a grand house in London. I am assured your soirées will be the toast of the town and your dance card will always be full, but I do hope that, despite the brilliant society, you will find the time to paint.

Even if Tom had not told me yesterday that his cousin Cecily is having her pictures exhibited at the Royal Academy, and had I not known full well that yours are every bit as good as hers—if not better—I would be saying this still. Never was any girl more suited to the elegant life than you. You have an eye for color and form and your talent should not be dismissed merely as a pleasant pastime for a young woman; it must be recognized as a true gift and not wasted, for this is part of what makes you, you. In less lofty terms, as Wonderland wisdom would have it, "don't lose your muchness."

The longer I am at University, the more firm my conviction becomes that it is vital to pursue one's passion. While my way is often hailed as being more scholarly and serious, there is no reason why the Universe should whisper more through Science than through Art. Regardless of the path we choose, I am starting to realize that strange and wonderful secrets creep out of the shadows for those who walk with wonder. I have come to agree with Maxwell that "scientific truth should be presented in different forms, and should be regarded as equally scientific, whether it appears in the robust form and the vivid colouring of a physical illustration, or in the tenuity and paleness of a symbolic expression."

You must excuse me a moment while I gather my scattered thoughts. I had left my table for supper and have only just returned. Perhaps this letter to you was in my mind, but as I walked back I was struck by the desire to show you a little of what I see every day, for I am convinced these environs are shaping the man I am becoming.

Cambridge is the most delightful place on Earth, Lizzy. I do not know how to describe what it means to me—except to say that it is the home of my soul. Tranquility is everywhere, from the refreshing green gardens to the softly flowing waters of the Cam. One of the reasons I love punting so much is that it mirrors how I feel about being here—like a wave in the river, flowing out of all that came before me, merging into all that is to come. This air has been consecrated by lofty thought, this ground hallowed with hours of honest toil, and all around there lingers a gentle grace. The centuries have left their mark: there are architectural delicacies to suit every palette; each college or chapel is a feast for the eyes; among the gothic spires, medieval arches, ancient pavements, and open courts, you can gaze upon intricate complexity or idyllic serenity, whatever most fits your thoughts at the moment. Yet for all its illustrious history, Cambridge shows its age only in mellow smiles and quiet wisdom; to all appearances, it is young and gay and as ever new as the Spring.

As you probably recall, when I set out for Trinity, the scientist I most venerated was Isaac Newton. His towering intellect cast its shadow on the centuries; he seemed the very picture of unapproachable perfection to me, and his marble statue in the college appeared a singularly befitting depiction of an idol. For the first few days, I was overwhelmed to be here at his alma mater, and often found myself wondering if—quite literally—my feet were falling where his invisible footsteps lay. So many months later, however, while I still look up to Newton as much as ever, I think

I have found a new mentor—one to whom I can relate on a personal level. James Clerk Maxwell walked these halls, too, and the warmth of his influence appeals more to me than Newton's cold perfection.

The study of Maxwell's electromagnetism occupies my days right now, but his presence in my life is more than just that. When I am at the Cavendish Laboratory, I can sense the affection and the careful consideration Maxwell put into designing it. He thought of this laboratory as being a new organ in the living body of the University. Top-notch equipment fills rooms flooded with light; the stairwells and corridors are clear for apparatus, and large spaces have been left empty to make room for all the experiments yet to come.

Maxwell placed a high value on the knowledge of dynamical principles, and believed that science education should involve both the intellect and the senses, for by so doing, "we shall not only extend our influence over a class of men who are not fond of cold abstractions, but, by opening at once all the gateways of knowledge, we shall ensure the association of the doctrines of science with those elementary sensations which form the obscure background of all our conscious thoughts, and which lend a vividness and relief to ideas."

Maxwell held that neither theoretical training nor experimental expertise on its own was sufficient, that the two must be reconciled in such a manner that we learn to recognize, in the world around us, material manifestations of the mathematical relations we study in books. The bridge between concrete and abstract must be crossed repeatedly, until it becomes so familiar that our thoughts automatically run in a scientific channel. It is as the Gryphon told Alice. When the Mock Turtle said they did lessons "ten hours the first day, nine the next, and so on," the Gryphon explained, "That's the reason they're called lessons . . . because they lessen from day to day."

I wish I had been born some years earlier so I could have arrived at Cambridge while Maxwell was still here, but he has left much of his soul behind. Sitting in the laboratory he so lovingly established, reading his many works, both scientific and philosophical, I have come to feel I know the man. Maxwell felt strongly that a University was a place for liberal education, and that even those of us who "make the pursuit of science the main business of our lives" must constantly endeavor to make connections between our work and other studies, "whether literary, philological, historical or philosophical." He warned against the "narrow professional spirit which may grow up among men of science," and the tendency of men to become "granulated into small worlds, which are all the more worldly for their very smallness."

You must see why these words bring you to mind. Remember the fight we had at Crystal Palace when we went to see the balloon ascent? While I was anxious to discover every aspect of the flight—the design of the delicate vessels, the manner of storing fuel, the dangers of wayward winds, the mechanics of descent—your gaze was fixed on the beauty and ornamentation of the balloons, which you said looked like floating Fabergé eggs. You grew tired of my incessant chatter, I became frustrated with your lack of apparent excitement, and we broke into quite a quarrel. But then the balloons took flight, and you reached out your hand to me and we stood together in rapt attention, letting wonder take over. I now begin to recognize, with the benefit of age and distance, that you and I complemented far more than we opposed each other. In my mental fencing with you, my thoughts grew stronger; even when I protested your views, they were enriching my own perspective, and I hope you will find that was true for you, too.

Maxwell believed that expressing ideas in words rather than symbols was an important part of the scientific process. In fact, he held that this should be the first step, so that before a student becomes embroiled in intricate mathematics, "the mere retention of which . . . materially interferes with further progress," he first comprehends the system he aims to describe. To this end, he constructed mental models and sought out analogies, often applying several of these to a single phenomenon; in each case, he pushed these analogies and models to their logical limits. He learnt both from what they were able to explain and from the places where they broke down.

It is only after thoughts are clarified that Maxwell advocates the use of mathematics, because then it adds to the picture instead of befuddling it. I know you are not overly fond of the subject, but if you were to leave your old prejudice aside for a minute and think about it, what is mathematics but the symbolic representation of relationships? Links and causes are illustrated succinctly by symbols and patterns that can be grasped at a glance; equations are more like paintings than poems because they simultaneously encompass an entirety instead of breaking it up into a sequence. As you learn to work with symbols, thought itself becomes economized. All the layers of connotation that shroud words fall away, and you are left simply with bare ideas, which can be manipulated in all the ways logic will allow. You must see, Lizzy, that there can be no language better suited than mathematics for the exploration of phenomena to which we know no parallel.

It is to Maxwell that you may ascribe the somewhat uncharacteristic reflections that make up this haphazard letter. Only this week, I read about an experiment he conducted on the nature of color. Even after Sir Isaac Newton demonstrated that white light was in fact a braided rainbow whose strands could be separated with a prism, much about light and color remained unknown. In one of his investigations, Maxwell attempted to build white from "blocks," as it were, of other colors. His method was rather ingenious, and surprisingly easy to replicate. He painted wooden tops in bright colors and spun them around so fast that the sharp boundaries blurred and the colors melted into each other.

This reminded me of the "wheels of life" we used to produce for parlor entertainment, a decade ago. Using the somewhat rickety machines of my devising, we would arrange a series of your beautiful little drawings on a spinning wheel, and when we set it in motion fast enough, the images blurred into each other, their transitions vanishing entirely, giving the appearance of movement. Even though we were intimately acquainted with every stroke, every cog—having put it all together—yet we would watch mesmerized.

Maxwell did something similar, except that he went one step further, pushing the tops into the stillness that lies beyond motion. He spun them so fast they appeared to stop. Maxwell tried various combinations before he found that, when the red of fire, the blue of water, and the green of earth meet in equal measure and whirl like mystic dervishes in Eastern lands, they throw off their individualities in a wild frenzy and merge into each other to become pure brilliant white.

I amuse myself with the fancy that in those magical moments when the trinity of red, blue, and green went into a hypnotic trance and emerged as white, a veil was drawn from Maxwell's eyes. Perhaps his enhanced vision enabled him to spy an underlying unity in apparently disparate objects, because a few years later he was able to see the phantoms of a beautiful and harmonious relationship that tied together electricity and magnetism—phenomena that had long been deemed distinct.

The story begins with electricity. I know you have been suffering my enraptured soliloquies on the subject since childhood, so I will attempt to restrain myself, but really, Lizzy, even you must admit how marvelous was the spectacle when we went to see Gilbert & Sullivan's new opera at the Savoy and found the candles replaced by electric lights. While, unlike me, you preferred being among books at Hatchards rather than at an exhibition of electromagnetic machines, the electric globe lights on Regent Street impressed us both equally.

I know you were less than pleased when Father and I spent hours engrossed in electrical experiments. For my own part, I was fascinated by the mere fact of currents and wires; for Father, I suppose the lure was to recreate his own childhood memories of the eloquent and dramatic Christmas lectures he heard Michael Faraday give at the Royal Institution.

I remember how fondly he recounted the time when Faraday sat, quite calm, inside a wooden cage, covered with metal, foil, and wire, and charged to about 100,000 volts. Faraday was secure in the knowledge that, since there was no electric field inside a charged conductor, he would be safe; people watched, convinced he would be killed, but while sparks flew everywhere outside the cage, science was borne out and he remained unharmed. Surely a child who witnesses a spectacle like that is not liable to forget it!

Perhaps it was for different reasons, but Faraday was quite a favorite with you, too. As soon as we heard the story of the bookbinder's apprentice, he became, in your eyes, the young hero of a modern-day science fairy tale. I must admit, it is quite a romantic story: A blacksmith's son was possessed by books night and day; by the light of day, he tended to the books, healing their scratches and tears; in the evenings, as he poured over their pages, the grateful books transferred all the knowledge they held into his ever-expanding mind. Thus passed seven years in honest labor and learning, until one day a customer gave young Michael tickets to a lecture by Sir Humphry Davy at the Royal Institution. This was, you insisted, the equivalent of Cinderella's Ball, for it was here that things transformed, as if by magic. Not only was his own mind set on fire by what he heard at the lecture, but the great Sir Humphry was so struck by the boy's enthusiasm that he took him on as an assistant.

Sir Humphry, of course, was quite a distinguished personage, and the opportunity to learn at his side was a huge boon to the young Faraday. Sir Humphry was glamorous and passionate, and a favorite of the metropolitan elite. Attracted by his masterful eloquence and his zeal, people of rank and fashion flocked to the lectures. What a figure he must have cut, in his kid-leather gloves, starched collar, and white cravat, speaking with passion about the rich beauties and sublime secrets of Nature. There are some who call him a dandy, because a large number of his audience were women; be that as it may, it is said that on days when Davy was speaking, the streets outside the Royal Institution were so crowded with carriages, it might as well have been a noon-day opera house. All that popularity cannot possibly be attributed to his looks! The truth is that Sir Humphry was a brilliant scientist who was remarkably committed to remaining accessible to the common man. Despite the rigorous demands of his own experimentation, he delivered a consistent stream of enthusiastic lectures full of spectacular demonstrations and genuine information. In fact, he even reshaped his own laboratory, in the basement of the Royal Institution, to accommodate spectators!

Even though Faraday had little formal training in science, his years with the books paid off; he had a feeling for the subject, and soon became a very successful experimentalist. Perhaps his apprenticeship with Sir Humphry taught Faraday more than just science; perhaps he also imbibed the ability to communicate ideas with clarity and ardor, and to accompany words with appropriate theatrical demonstrations, for he too became an extremely popular and masterful lecturer.

I shall not, however, presume too much upon your old fondness for Faraday's history, and will refrain from pestering you again about electricity, other than to remind you that there are two types of charges, called by convention positive and negative; that like charges repel each other whereas unlike charges attract; and that an electric current is but the flow of charge.

Since the story involves both electricity and magnetism, I need to rekindle your memory on one other count. I can almost hear your exasperation, so let me say that I do not honestly think you can have forgotten the fond hours we spent with Father's compass, provoking the hapless needle to turn now here, now there, in an endless circular chase. We were fascinated by the way the magnet could move the needle without ever touching it. We were so young then; the only ways we knew to move things were to physically tug away or push with all our might. This mysterious force seemed wasted in the aimless wanderings of a needle when it would have been so much better employed in moving chairs or houses or trees or people. That, I am sure you recall. All I wanted to remind you of was that the compass needle also moved in the presence of an electric current. It appeared that the flow of electric charge mimicked the behavior of a magnet, but we were never quite able to puzzle out why.

Father did not know either, when we asked him the whys and wherefores of this duplicitous behavior. All he had to offer was another Faraday epigram: "The beauty of electricity or of any other force is not that the power is mysterious, and unexpected, . . . but that it is under *law.*" To my nine-year-old mind in search of a mechanical explanation, a neat quotation was not a satisfactory substitute. I was disappointed by Father's uncharacteristic, imprecise reply, and the ability of a current to deflect the compass needle continued to bother me. Now, after all these years, when I finally begin to understand this phenomenon, I feel that even poetry would scarcely do it justice.

But before I share this beauty with you, I must tell you a little about electric and magnetic fields. The concept of a field was one Maxwell inherited from Faraday. In order to explain how two magnets could exert a force on each other without touching, Faraday postulated the existence of what he called "lines of force." These lines, so distinctly visualizable, existed in the seemingly empty space between the two objects; the distribution of the field was reflected in the pattern of these lines, through which the force was mediated. Maxwell realized the intrinsic elegance of the construction and marveled at how Faraday's "lines of force can 'weave a web across the sky,' and lead the stars in their courses without any necessarily immediate connection with the objects of their attraction."

He took Faraday's intuitive idea and made it quantitative. Since the lines of force represent the field, he said, the density of lines in any region of space should provide a measure of the field strength there. Using his equations, Maxwell was able to manipulate the field in ways that lay beyond the scope of Faraday's pictorial representation. Like Newton before him, Maxwell harnessed the power of mathematics to express abstract relationships precisely, to build on the theory and extract concrete predictions from it.

Faraday himself was surprised at the outcome. He later wrote to Maxwell, "I was at first almost frightened when I saw such mathematical force made to bear upon the subject, and then wondered to see that the subject stood it so well." This was one of those occasions when translating the concept into mathematics was not merely an empty academic flourish but a crucial step that led eventually to a brilliant insight.

The basic idea is simple enough: that each electric charge carries with it an invisible net that it throws out into space. This net, which is thickest close to the charge and progressively thinner as it moves outward, represents the electric field. (An identical description can be made for a magnetic field, as the net cast by either of the two poles of a magnet, and all the arguments I will make for the electric field carry over to the magnetic field.) The field is not a tangible thing, but an intricate lace of messages written in a language known only to electric charges. The messages are clear instructions that tell other charges how to move—exactly how far and in which direction. These directives exist at every point in space, whether or not a charge is present to implement them. Any charge, when placed in an electric field, will read the directions inscribed at the point where it finds itself, and act accordingly; as a result, one charge seems to "push" another, even though the other is far away.

By a previously agreed-upon convention, the electric field addresses its messages to a single positive charge. I picture these as being the simplest creatures, unable to decode anything but the most direct instruction; all other charges must do some basic arithmetic to find out what instruction is meant for them, and of this operation they are apparently able. An object that carries

ONLY THE LONGEST THREADS

three positive charges must implement the basic message thrice over; but one can hardly triple a direction, so the object merely has to move three times as far in the prescribed direction. Negative charges, on the other hand, are contrary creatures; an object carrying five negative charges will travel five times as far as the basic message says, but in exactly the opposite direction!

There is one complication I have omitted in this discussion: the fact that every charge we bring in casts its own net, which spreads through space as a rippling wave, its strands interweaving seamlessly with the other nets already present. An innocent wayfaring charge that wanders, deliberately or unwittingly, into this silken web, must now obey the dictates of the composite field, which takes into account all the decrees mandated by each of the charges. Of course, the wayfarer comes enshrouded in a net of its own . . . and so the situation can get infinitely complicated, as is true of any party where every guest insists on talking! Sometimes, for ease of conversation, one talks of "test" particles, which are charges that can hear the ambient field without feeling compelled to add their own voice to the hum; but these are just mental tools we use to simplify the discussion— no such blissfully silent charge actually exists.

Even though thus far I have mentioned only electric fields, the selfsame discussion holds for magnetic fields. Since the Earth itself acts like a magnet, we are always in the presence of an ambient magnetic field, whether or not a visible source is present. That, in fact, is how generations of sailors have navigated the seas, by trusting magnetic compass needles to heed the call of the Earth's field and align themselves accordingly. If, however, there is another magnet closer at hand, sending out louder instructions, the faint calls of the True North are drowned out and the compass needle takes its marching orders from this new Drill Sergeant.

Lest you think these ideas sound implausible, let me tell you that it is these very fields that underlie the magic of wireless telegraphs! I remember how fascinated you were by the fact that words could be just plucked out of the air and materialize on paper! While outwardly I scoffed at your calling this magic, these missives sent across bare space, with no wires needed to carry them, entranced me too. For a while I was convinced that these messages only *seemed* invisible, that if I glared hard enough at the air around me I would see faint words flying around, like ghosts of carrier pigeons, heading straight to the nearest telegraph office. I never saw those apparitions, of course, and now I know why.

This has been a long digression, I know, but as you will see, it is important that you understand what a field is. This idea is what enabled Maxwell to reveal the deep, profound, and largely unsuspected connection between electricity and magnetism. Other than the fact that both these phenomena allow two types of charges, which obey the adage "opposites attract, and like repels like," there really is no apparent similarity between electricity and magnetism. In fact, they were long thought of as intrinsically different qualities, as distinct from each other as red is from blue. Gradually, over this past century, the realization dawned that the disparity was not quite as fundamental as we had thought.

The deflection of compass needles in the presence of current-carrying wires made it clear that, at the very least, electricity and magnetism were not completely indifferent to each other. This gave rise to a spate of investigations, particularly when it was observed that two current-carrying wires attracted or repelled each other just as magnets would have done, the "polarity" of the wires being determined by the direction of current flow. A static charge had no such effect, but the flow of charge seemed to mimic magnetism.

A decade or so of furious experimentation led to the conclusion that this phenomenon could also be "reversed"; the honor of producing electricity from magnetism fell to our friend Faraday, who demonstrated that a changing magnetic field induced a current to flow in a nearby circuit. It was obvious that we had stumbled upon some fundamental connection between electricity and magnetism, but the precise interpretation was not immediately clear.

This was the scene when Maxwell stepped onto the stage, and he put up a show the likes of which the Music Halls have never seen. Not ventriloquists nor tightrope walkers, not sword swallowers nor fire princes, could compete with the magic Maxwell conjured. Invoking the field, much as magicians of old waved their wands, he magically made sense of everything. Under his deft strokes, all the scattered observations made about these phenomena rearranged themselves into a neat little set of equations which laid out how an electric field can result from changes in a magnetic field. These equations sketch, in their own mathematical shorthand, the following picture.

Consider, for the moment, a field that consists purely of messages addressed to magnets. If the message at every point is fixed, an electric charge that strays into the fold will be blissfully oblivious to the instructions floating in the air around it. But, should these messages change, the electric charge is affected. This is quite a curious fact. After all, if you were impervious to certain aspects of your surroundings, it would be quite logical to assume that you would be indifferent to how they changed; yet that is not so with electric charges. While they completely ignore the constant hum of magnetic messages, regardless of how loud and strong these might be, the instant there is a change in the ambient magnetism, electric charges prick their ears and listen.

Now, it turns out that there is no room for eavesdroppers in Nature, and what one hears, one must obey; hence electric charges, which completely brush off the whispers of a constant magnetic field, find themselves constrained to obey certain dictates in a changing magnetic field. In fact, judging from their behavior, they might as well be in the presence of an electric field.

The situation was strikingly symmetric to something we had observed earlier: the generation of a magnetic field by a current. If that statement is reframed purely in terms of fields, we find ourselves saying that the changing electric field caused by a flow of charge (a current) gives rise to a magnetic field. As he wrote down the equations describing each of these processes, Maxwell noticed that they were almost mirror images of each other. Put together, they were more potent than anyone, even Maxwell, had suspected!

So that you may truly appreciate the next chapter of this story, pause and reflect for a moment on what happens when a field changes and the "messages" encoded across space are no longer static. Let us consider a particular kind of variation—one where the electric charge sourcing the field bobs up and down. As in a rope with an oscillating endpoint, waves are set up in each of the force lines tethered to this charge. These periodic disturbances ripple throughout the net, causing the "instructions" to fluctuate. Standing still, at any given point, one would experience these messages getting progressively stronger and demanding an increasingly dramatic response, until they built up to a crescendo and then, exhausted, became ever weaker until they returned to their initial state, where the journey would begin again.

As he contemplated these matters, Maxwell came to a startling conclusion: oscillatory behavior calls oscillatory behavior into being. Suppose a magnetic field changes, not by arbitrary or unbounded amounts, but in a controlled, periodic manner. The ebb and flow of the magnetic field is reflected in the fluctuations of the electric field generated by this motion; an undulating electric field calls forth from the void a magnetic field that vibrates in sympathy; and so the circle keeps turning, ad infinitum.

Etched into his equations, Maxwell found the ghosts of these untiring oscillations, closing in time and again, only to move apart once more. Electric and magnetic fields move to and fro, like an endless row of alternating pendulums, perfectly synchronized in perpetual motion. The exercise was orderly, even beautiful, but after a while it began to seem quite meaningless— until Maxwell realized the deep subtlety of this performance.

Focusing merely on the repetition was like reducing pure art to calisthenics. The synchronized motion of the fields was akin to the seamless movements of the corps de ballet—pleasing, but not the high point of the show. As his inspired gaze pierced the swirl, Maxwell noticed that, as the members of the troop wove back and forth, something changed hands, something that was being pushed forward by each oscillation. Light.

Light was the true prima donna. The whole troop, the entire dance, was in her service. All those tightly choreographed, ceaseless motions suddenly made sense. They were there to propel her forward. Electric and magnetic fields oscillate untiringly because, in doing so, they propagate light.

On the surface, Maxwell's equations laid out the manner in which an oscillating electric field generates an oscillating magnetic field, and vice versa. But it turns out that this eternal rocking back and forth forms a cradle for light. Light is embedded in Maxwell's theory as a wave, a self-sustaining undulating disturbance that spreads through the electromagnetic field at a fixed speed, dictated by the equations.

Such is the story behind the dim, wavering light of the flame under which I write this letter, and the light—perhaps the bright, straight beams of the sun—under which you will read it; wondrous, is it not? But even more wondrous, at least to me, is the fact that there is such a deep common thread running through such apparently diverse phenomena. Maxwell has, probably for the first time, come up with a scheme that subsumes two previously separate entities, uniting them into one coherent, consistent whole.

This feat is so spectacular, one cannot help but try to retrace Maxwell's steps, to puzzle out how it happened. It is the same instinct that causes your eyes to dart around, looking for strings and trapdoors, when the magician's assistant disappears in a puff of purple smoke. Maxwell was not a trickster, and he felt no need to hide his methods. The true power, he knew, lay not with him, but in Mathematics. Even when we perform the manipulations ourselves, he wrote, "the equivalence of these different forms, though a necessary consequence of self-evident axioms, is not always, to our minds, self-evident."

When we analyze systems mathematically, we begin to notice the recurrence of certain mathematical forms in situations that have no apparent similarity. In such cases, though the physical interpretation of the quantities may differ widely, "the mathematical forms of the relations . . . are the same." As a result, the "trains of reasoning" resemble each other so much that we may proceed by mathematical parable, using our knowledge of one system to gain insight into the study of the other. If we had limited our exploration purely to the physical aspects of these systems, we would have found no similarity whatsoever; the resemblance appears only in the mathematical descriptions of their internal relationships. It is this similarity of form that caused Maxwell to draw an analogy between electricity and magnetism, revealing connections that would otherwise have stayed hidden.

I am extremely intrigued by this concept and eager to know if the scheme can be extended to include other natural phenomena as well. How wonderful it would be to uncover the hidden threads that tie this varied world together, connecting us all like intricate lace! That thought put me in mind of you, and your belief that even apparently incompatible personalities have much in common; it is just hidden from view. Maybe, instead of focusing on our differences, we should be searching for these invisible linkages between Men. Perhaps these similarities are what will hold us together, in peace, and as one.

My eyes are so heavy with sleep that I begin to wonder if my writings are fully coherent. I eagerly await Christmas break, when I can visit home and hear all your news in person. I want to take those long walks with you one last time, before you go off on your husband's arm into "that new world that is the old."

The old clock strikes again, loudly. The nightingale perched outside my window flies off, in protest. Her lilting song has been drowned out one time too many. I agree with her that the chimes can no longer be ignored. I suppose I should heed their message and turn in.

Now I must be off to bed, and lest I grow too nostalgic about your impending marriage, I shall think only on the endless supply of Fortnum and Mason preserves that will no doubt fill your larders; the tickets to Friday Night discourses at the Royal Institution that Julian will ply me with; and all the enchanting young ladies you will shortly be introducing to

Your dashing and brilliant younger brother,

Charles.

Email: Sara to Leo

From: Sara Byrne <breaking.symmetries@gmail.com>
Date: Tue, Oct 30, 2012 at 6:45 PM
Subject: Yay!
To: Leonardo.Santorini@gmail.com

Dear Leo,

Yay! You're really doing this! I hoped the protracted silence didn't mean you had second thoughts. I'm so glad that wasn't the case.

I wanted to savor your manuscript, so I printed it out and headed to the most peaceful and quiet place I know. In the stacks at Widener Library, towering bookcases seem to shut out external distractions, and the whole world shrinks to the span between two shelves. The faint, comforting smell of old paper was all that remained of my surroundings, as your words transported me to a different place and time.

I must admit I was a little taken aback by the beginning. It wasn't just a few unfamiliar words that threw me off; even the rhythm of thought felt foreign. But slowly I began to get a sense for the attitudes that prevailed then, and I could see how iconoclastic Newton's methods must have appeared to eyes that were still clouded by the occult. I realized how little I knew about the other scientists of that time, so I decided to find out more, at least about the members of the Royal Society.

One of the first items that showed up in my Google search was Bacon's argument about the need to change the logic of the sciences, to leave behind the confusion of syllogisms and proceed using induction. His words are so impassioned, their power resounds through the centuries. How grand he sounds when he says, "I do not propose merely to survey these regions in my mind, like an augur taking auspices, but to enter them like a general who means to take possession"! Why don't we talk like that anymore?

I love that you chose to see electromagnetism through an artistic lens. That really lets the beauty of the theory shine through. Reading about Maxwell's legendary equations in such unfamiliar language, I was struck again by the creative tension between the objectivity of facts and the subjectivity with which we internalize them; each of us cloaks the world with the mantles of our own experiences.

Charles's descriptions of his environs were so evocative, I wanted to hop on the next flight to London. I knew from your email that the Royal Institution has a museum, but apparently they also continue with Friday Night Discourses and the annual Christmas lectures, in the same hall where Faraday once lectured. Bless England for hanging on to tradition! I've put both those things on my bucket list.

There was a vivid sense of place about that letter. Cambridge seemed integral to the story, not incidental. It was as if it were a physical conduit for thought through the ages, or a sort of touchstone for reason. That made me wonder if our intellect is shaped by the physical spaces we inhabit. As I walked back to the Physics Department, I took a good look around Harvard Yard. Are my impressions of physics any different for having been formed here? In the years to come, when I move on, will the flavor of this place linger and color my understanding?

Enough philosophizing. I should get lunch before I keel over from hunger. Maybe I'll grab a sandwich and go eat it outside. Leaves freckle the Yard. Red as bricks, they crackle and crumble under busy feet. There is a delicious, bracing nip in the air. The fall makes me so happy. Hope you're enjoying it too.

Best,

Sara

P.S. Thank you for introducing me to Maxwell. I had no idea he was so cultured and engaging. His writings are an absolute delight! The more I read, the more I want to know. I think I have a bit of a crush on him . . .

The Second Installment

From: Leonardo Santorini <leonardo.santorini@gmail.com>
Date: Tue, Jan 22, 2013 at 3:09 AM
Subject: The Second Installment
To: breaking.symmetries@gmail.com

Dear Sara,

I'm under a deadline at work, but while my mind is in a different century, I can't focus on anything in the here and now. The only way I will ever stop fidgeting with these chapters is to get them off my hands. So, here they are.

The third chapter is set in New York City, in the wake of World War I. The narrator is a thirty-something schoolteacher who is exceedingly enamored with the esoteric theory of relativity, and its author. To Einstein, the grand aim of science was to explain the maximal number of facts using minimal assumptions, and he excelled at this art. A master of unification, Einstein stitched together time and space, and related mass and energy. As was inevitable, he tried to broker a marriage between light and gravity, the protagonists of special and general relativity; but he failed. His stubborn denial of quantum mechanics was partly to blame. A theory formulated in terms of probabilities could not, he thought, "bring us any closer to the secret of the Old One."

Einstein's inability to accomplish his cherished goal is portrayed as a fallibility of genius. We paint sad pictures of a white-haired maestro singing what he called his "lonely old song" until the end. If only he had been less obstinate, we sigh, as we lament the loss of what he could have accomplished if he had relaxed his objections. Until I started writing this book, I felt the same. I know better now. If Einstein had been able to treat his work with the necessary detachment, he would have been a lesser scientist, not a greater one. It takes all the faith, commitment, and passion you can muster to build a universe in your mind. The relationship between scientists and their ideas is intense, emotional, obsessive, demanding. Ideas are not commodities, they are living things. You cannot choose which you fall in love with; one is not interchangeable with another. In science, as in life, you must be true to yourself if you are ever to accomplish anything of value.

Rejecting the myth of the dispassionate scientist as being "not only contrary to experience, but logically inconceivable," Michael Polanyi makes an excellent point: "Courts of law employ two separate lawyers to argue opposite pleas, because it is only by a passionate commitment to a particular view that the imagination can discover the evidence that supports it." This is how it must be done. There is no other way. Subjectivity does not detract from science, but on the contrary enhances it. The beauty of having so many different minds focused on a task is that no single person, regardless of his or her stature, is responsible for the whole. Each intelligence is free to pursue its own passion. Truth emerges naturally out of the collective.

The fourth chapter is narrated by a postdoctoral researcher, at Bohr's institute in Copenhagen, in the early 1930s. Quantum mechanics was slowly gaining acceptance. The new generation was willing to meddle with ideas that seemed sacrosanct and immutable to their elders. They surrendered their claim to certainty and learned to navigate a world that could be described neither by words nor by images. The framework of physics grew by leaps and bounds, but there was a lot of heartbreak along the way. Unable to resolve quantum conundrums, many brilliant scientists fell prey to depression. Some, like Paul Ehrenfest, even gave in to suicide. At this time of revolution and unrest, Bohr's institute became a refuge for physicists who craved the camaraderie of conversation.

As always, I eagerly await your feedback. I'm particularly curious to hear your thoughts on the quantum mechanics chapter. I wanted this to stand apart from the diaries and letters of the classical era, so I put a different spin on it. I'm not sure how that comes across, so please let me know what you think. If something doesn't make sense to you, odds are it won't make sense to anyone else either.

When you have time, do fill me in on your news. How's your work going? Are you enjoying the winter? Have you given any serious thought to writing that string theory chapter?

Ciao,

Leo

P.S. One way to make your writing clearer and more targeted is to address it to a specific person—your "ideal reader." The tactic is a common one, but I had never seen it used in a science book, until I read Einstein and Infeld's *The Evolution of Physics*. Not only do these authors have an ideal reader in mind, they describe him rather vividly. His great interest in ideas compensates for his lack of concrete knowledge, they say, and they admire the patience with which he tackles the more dry and difficult passages. I get the impression this imaginary character was quite real to them. I bet they could even picture him.

CHAPTER 3

A Holy Curiosity

[The Special and General Theories of Relativity]

Never lose a holy curiosity. —ALBERT EINSTEIN

APRIL 3, 1921
NEW YORK

It finally happened. He arrived yesterday afternoon. The *Times* headline this morning screams:

PROF EINSTEIN HERE. EXPLAINS RELATIVITY.

'Poet in Science' Says It Is a Theory of Space and Time.

BUT IT BAFFLES REPORTERS.

I hardly slept last night, so anxious was I to see the morning paper, to read what people had written about this monumental event, to see the photographs, and to reassure myself that it was not all just a dream. For among those "thousands of spectators . . . in Battery Park when the mayor and other dignitaries brought Einstein ashore on a police tugboat," I was also present.

With uncharacteristic assertiveness, I had propelled my way to the front of the singing, flag-waving masses awaiting the arrival of the steamship *Rotterdam*. A few people glared at me, but I was only dimly conscious of the disapproval they aimed at the short stocky man with thick glasses so desperately tunneling his way through the crowd. All I knew at that moment was that I had to see Einstein, the man who has unknowingly been my mentor this past year. How much he has come to mean to me!

I first heard of Einstein when he burst upon the international stage in November 1919, after the now-famous results of the Eddington expedition were announced in London. One of the concrete predictions of Einstein's general theory of relativity was that, in the presence of a gravitational field, light no longer travels in perfectly straight lines. This led him to predict that starlight would be deflected as it grazed the sun. As Arthur Eddington's photographs proved, Einstein's calculations agreed exactly with observation. The President of the Royal Society declared this to be "one of the most momentous, if not the most momentous, pronouncements of human thought," and the *Times* of London reported that there had been a

REVOLUTION IN SCIENCE.

NEW THEORY OF THE UNIVERSE.

NEWTONIAN IDEAS OVERTHROWN.

But despite the compelling testimony of bent starlight, the prevalent attitude toward Einstein's theory was one of confusion. The consensus was that no one had yet succeeded in stating it in clear language. The *New York Times* further strengthened this impression in its own breathless headline about Eddington's observations:

LIGHTS ALL ASKEW IN THE HEAVENS

Men of Science More or Less Agog Over Results
of Eclipse Observations.

EINSTEIN THEORY TRIUMPHS

Stars Not Where They Seemed or Were Calculated
to be, but Nobody Need Worry.

A BOOK FOR 12 WISE MEN

No More in All the World Could Comprehend It, Said
Einstein When His Daring Publishers Accepted It.

That clinched it for me. My curiosity piqued, I decided to find out more about this scientist-celebrity. One of the first things I discovered was that acclaim had not come to him overnight. Most of the work being so enthusiastically eulogized had actually been carried out more than a decade ago. What we were celebrating was the proof of his visions.

As I read some more, I found out that there was not just a general theory, but also a special theory of relativity, which had come before it. Both these theories, which question all our preconceptions and turn everyday logic on its head, were conjured up by the power of pure thought, by a clerk who worked by day in a Swiss patent office. The fact that such a thing is possible was absolutely thrilling to me.

Much as I love teaching my high school students, it is not always intellectually satisfying. For years I regretted not having the opportunity to go on to a university. My parents worked as hard as they could, but resources were scarce. Both my sister and I dreamed of a better life. Esther fantasized about marrying a rich man and becoming an elegant lady, with lace gloves, flamboyant satin coats, and ropes of pearls—like the beautiful women who come to the perfume counter at Macy's, she would say. Outwardly, I scoffed at her dreams, telling her that she was being materialistic, that tinsels and gold did not endure, but I had secret ambitions, too: I was to become a famous scientist, remembered down the ages for my contributions.

From the histories I devoured at the local library, I had developed quite a romantic notion of the underprivileged boy who triumphs over circumstance to become a scientific legend: Isaac Newton went to Cambridge as a sizar—little more than a servant to boys who were inferior to him in all but worldly status. Samuel Pepys was a tailor's son. Michael Faraday was a bookbinder's apprentice. Feeding myself these stories, I had become convinced that Life deals you a bad hand in the beginning only so she can make up for it later, but that childish faith was not borne out, at least for me. Father died in an accident when I was only eighteen, and my dreams died with him. I was no longer at leisure to be a poor student; I had to get a job to help support the family. Over the years I suppose I became somewhat embittered about this—until I discovered Albert Einstein.

Bernard Shaw said that where Napoleon and other great men made empires, Einstein made a universe. When I read that, it brought back some of the assurance I had felt in childhood; it reminded me that the dominion of the intellect is unbounded, and open to us all. My notions are not as naive as they were ten years ago, but now I do not feel caged in by my job either. Einstein has proved that ideas do not restrict themselves to the air around academic institutions, and that neither your station nor your occupation need keep you from studying what you love. The realization was intensely liberating. I had long since yearned for the chance to flex my mental muscles with a real intellectual challenge, so I decided to try my hand at learning the world-changing theory constructed by Einstein as he worked as a clerk in a Swiss patent office. I looked around for a suitable text and found that a surprising amount had been written on the subject.

During my search, I chanced upon a delightful little volume by Edwin Emery Slosson entitled *Easy Lessons in Einstein*, which seemed to be intelligible. What particularly piqued my interest was the manner in which the book starts. In place of a conventional Introduction was a Dialogue, declared to be a preventative measure against any prospective reader buying the book under false pretenses. This Dialogue, between the Reader and the Author, takes place in a streetcar as the Reader, glancing at the newspaper, remarks how queer it is that an entire page is devoted to the discussion of a discovery in physics. He reads out some of the sentence fragments, which are every bit as sensational as the recent headlines. Relativity is hailed as the most sensational discovery in the history of science, the greatest achievement of the human intellect, etc.

"It looks as though I ought to know something about this, doesn't it?" remarks the Reader, and the Author agrees that this is probably something he will have to come to grips with, sooner or later. The Reader continues his perusal of the column, marveling at the fantastical nature of the claims—"Parallel lines meet"; "a man moving with the speed of light never grows old"; "gravitation due to a warp in space"; "length of a measuring stick depends upon direction of its motion"; "mass is latent energy"; "time as a fourth dimension"—and concludes that Einstein must be crazy. Here the Author dissents. There must be some method to Einstein's madness, he says; "otherwise how could he have hit upon the exact extent of the sun's attraction on light?"

Recognizing the superior knowledge of the Author, the Reader asks if he will explain, in simple words, what relativity is about, and the Author says he will do "just that. I can tell you what it is *about*, though I can't tell you what it *is*." He explains that, without plunging into mathematics, he can talk about some interesting aspects of the theory, and the Reader can pursue the subject more seriously later, if he likes. The Reader takes the Author up on his offer, and thus begins the book. Being in much the same frame of mind as the Reader, this book struck me as the perfect stepping-stone to my own study of the subject, and I was not proved wrong. Slosson's explanations were both engaging and clear. He helped build up my confidence and left me wanting to learn more. So, after working my way through this volume, I turned to Einstein's own book entitled *Relativity: The Special and General Theory*.

It was when I read his book that Einstein truly came alive for me. As I stood at Battery Park yesterday, waiting for a glimpse of the great man, I thought of how often I had heard his voice in my head. The very first lines drew me in. "In your schooldays, most of you who read this book made acquaintance with the noble building of Euclid's geometry, and you remember—perhaps with more respect than love—the magnificent structure, on the lofty staircase of which you were chased about for uncounted hours by conscientious teachers." He wrote so intimately, I immediately settled in to what felt like a conversation with a friend, not a lecture by a "scientific mind whose deductions have staggered the ablest intellects of Europe," as the newspapers proclaim.

Fascinated, I tagged along as Einstein walked me through the concept of truth (as being something that follows logically and consistently from a set of axioms) and convinced me that to talk of bodies changing their position in space with time, without "serious reflection and detailed explanations," would be tantamount to sin, since "it is not clear what is to be understood by 'position' and 'space.'" Thus gently, he nudged me to probe the calm waters of the seemingly simplest concepts and examine the many layers of unstated assumptions that lay underneath.

In the very early years of this century, when Einstein turned deep conundrums over in his head on his walk home through the picturesque streets of Bern, light was very much on his mind. Light had turned out to be quite a problem child, despite the love and attention Maxwell had showered on it. This time, its victim was the deeply cherished law of Galilean relativity.

Centuries ago, Newton laid down a decree that I pass on to my own students every year. He ordained that every body will continue in a state of rest, or of uniform motion in a straight line, unless compelled by an external force to do otherwise; this, Newton's first law of three, is known as the law of inertia. Inertia is the tendency of a body to resist change and to continue in the future as it has in the past. Consequently, those of us who observe the world from platforms that are either uniformly moving or still are called inertial observers.

Newton argued that two inertial observers should perceive the laws of physics as being the same. In other words, nature should not appear to be fundamentally different to me, standing here, and you, standing five feet to my left. If both of us throw identical balls in the air, at exactly the same speed and in the same direction, they will land identical distances away from us. Nor should we disagree if one of us hops on a railway carriage that glides smoothly along the tracks, never speeding up or turning or slowing down. Indeed, if the carriage had no windows to the outside, and one could not see the landscape whizzing by, one might never realize that the train was moving at all. And if an inertial observer can never say for sure whether he or she is moving, then it stands to reason that the basic laws of physics remain unchanged under uniform constant motion. This eminently reasonable statement is called the principle of (Galilean) relativity and was held to be so obvious as to require no justification.

The arrangement worked for centuries while natural phenomena were explained in terms of classical mechanics. With the advent of Maxwell's equations, however, trouble started brewing. It is easy to see why: an electric charge at rest generates only an electric field, but to an observer in motion, this same static charge appears to be moving and thus must also give rise to a magnetic field. Which reality is true? Is the magnetic field actually there or not? Such paradoxes made it clear that the new theory of electromagnetism refused to comply with the age-old principles.

What played out in Einstein's head was nothing less than an epic confrontation between Maxwell's spectacularly successful electromagnetism and the long-revered framework of classical mechanics. As these theories were pitted against each other, many long-held, deeply rooted beliefs went down in flames; and out of these ashes rose the phoenix of special relativity.

While I cannot claim to completely understand the mathematical derivation of the theory, the intuitive idea does make sense. I suppose it is rendered accessible because Einstein thinks in pictures. He uses equations to work out exact relationships, and to frame his ideas with precision, but the flash of genius is in the mental image.

Einstein starts by making this abstract conceptual clash concrete. He imagines a railway carriage traveling along the tracks at a constant speed and direction, changing its position with respect to the stationary platform at the train station, yet doing so without rotating. I have become rather attached to this hypothetical carriage, so often does it appear throughout his book. To complete the setup of the mental experiment, he places a man on this carriage (I will name him David) while his friend (whom I name Marcel) stands on the platform. Since both David and Marcel are inertial observers, they should see natural phenomena run their course according to the same general laws. The numerical values they assign to quantities such as velocity and distance may differ, but the two must agree on the laws of physics.

Now Einstein takes a tiny step into complexity. Imagine a raven, he says. I found this quite reassuring; there is nothing remotely mathematical about a raven. We set this bird flying through the air in such a manner that its motion is uniform and perfectly parallel to the railway carriage. As expected, Marcel sees the raven as flying faster than does David; this must be the case, since the latter is himself moving in the same direction as the raven.

There should be a way to take the observations made by one friend and predict what the other will see, and indeed there is: basic addition. To make sure I understood exactly what Einstein was saying, I assigned concrete values to the speeds. I ran my mental train along the tracks at 50 kilometers/second and let the raven fly such that its speed, as measured by Marcel, is 70 kilometers/second. It is trivial to see that, in the first second, the raven covers a distance of 70 kilometers and the train moves ahead by only 50 kilometers. The distance between them increases by 20 kilometers, so David, who moves with the train, measures the speed of the raven to be 20 kilometers/second. This seemed simple enough. Thus far, there had been no surprises, and I began to settle more comfortably into the book. But perhaps I had judged a little too early.

So much for ravens, Einstein says; what now of light? Suppose that Marcel shoots a ray of light in the direction of the train.* He will then see light travel at the speed c, while David, using the reasoning above, should see light travel at the speed 50 kilometers/second less than c. But that can simply not be true. Light always travels at the speed c; this is a general law of nature, derivable from Maxwell's equations, and like every other general law must hold in every inertial frame of reference.

Rather abruptly, we find ourselves up against a wall. Our familiar logic does not apply to light. Something is clearly wrong; either Maxwell's equations are lacking, or else there is something amiss with the very sensible laws of classical mechanics. Faced with this choice, the traditional approach might have been to reject the new and hold on to laws borne out by years of everyday experience. Einstein, however, bet on Maxwell. This was not a rebellious, contrarian choice, but one motivated by the depth and beauty of the breathtakingly symmetric equations that explained not only how things happen, but why. Maxwell's theory was aesthetically and logically far more complete than Newton's laws of classical mechanics, which, though monumentally successful, were just working prescriptions.

As Dr. Slosson says, "when we get contradictory answers to the questions we put to Nature we must assume—unless Nature is nonsensical—that we are asking nonsensical questions. If in the trial of a pickpocket one witness swears that the thief did not run up the street and another witness that he did not run down the street the lawyer does not necessarily say that one of them must be a liar. He meditates a moment and then it occurs to him that possibly the pickpocket did not move or that perhaps he disappeared into the third dimension by climbing up a fire-escape or dropping into a coal hole."

This was what Einstein did. He reshaped our conceptions to accommodate a new possibility. Since he was questioning age-

---

*From now on, we assume for simplicity that the entire arrangement takes place within a vacuum, where the speed of light is c. In other media, though still fixed and calculable from Maxwell's equations, the speed is slightly slower.

old beliefs, Einstein had to construct a new vision of the world from the ground up. His building material was logic, but even so he needed a place to start. Some axioms were needed. As postulates, he chose, first, that the laws of physics should appear the same to all inertial observers, and second, the experimentally demonstrated fact* that the speed of light is always constant and appears to be invariant, regardless of the motion of the observer. These were the only foundations Einstein had. Making no additional assumptions, and setting all preconceptions aside, he set about examining what had to be true for these two statements to be consistent.

Velocity was the key; it is what distinguishes, and links, different inertial frames to each other. So Einstein started carefully rethinking the way in which velocity is measured. It is a composite concept, this ratio between distance and time. In order to compute the velocity of an object, we measure the distance it travels and divide this by the time taken to execute the motion.

Having introduced this concept, Einstein takes us back to the railway carriage setup, and examining exactly how each friend would go about measuring the velocity of light. Suppose David (the man in the carriage) aims a ray of light at the window, at right angles to the motion of the train. He can measure the distance between himself and the window and, equally, he can measure the time light takes to hit the window and be reflected back to him. Dividing the one by the other should give him the speed of light, the absolute value of which is fixed and known.

Now step into Marcel's shoes. He, too, sees a ray of light shot at the window, but as a result of the train's motion he also sees a change in the position of his friend. Marcel knows that if the ray of light were to simply reverse its path, it would end up back at the point where David *was*; it must travel at an angle, aiming ahead, to get to the point where David has been carried by the train. In doing so, this beam will cover a greater distance than

*Confirmed by the 1887 experiment of Michelson and Morley.

is apparent to the unsuspecting David, even though it is clear to Marcel. He measures the distance, which turns out to be simply the hypotenuse of a right triangle whose sides are given by (a) the distance between David and the window and (b) the distance covered by the train in the time taken for the ray of light to return to David. Marcel can now divide the distance traveled by the time taken to find the velocity of light. Being an inertial observer, who can claim full validity for his point of view, it makes sense that Marcel's calculations should yield the same value as David's. How is this possible?

As is familiar to schoolchildren everywhere, it is not difficult to manipulate your way to the answer if you know what it is! This is essentially what Einstein did. In a move of breathtaking audacity, he said that the only way for the two friends to arrive at a consensus regarding the speed of light is for Marcel (who perceives light as having traveled a greater distance) to see this process unfold over a longer time interval. So, as far as Marcel is concerned, the clocks on the train are running in slow motion! This somewhat unexpected resolution of the conundrum was just what the doctor ordered.

Time appears to be dilated, Einstein said, when an object is in motion. Since one is never in motion with respect to oneself, this holds only for other people's perceptions of my motion, or my perception of theirs. In my own frame of reference, things go along just as they always have. But if I look at people who are in motion with respect to where I stand, I will see their time as passing more slowly than my own. Paradoxically, since they are equally justified in claiming themselves to be still while I am in motion, they will see their own clocks keeping time as usual and perceive *mine* as being the one that has slowed down! Between these apparently contradictory realities, there is no objective way of deciding who is right and who is wrong; both are equally correct. Even when you have followed all the logical steps, this inescapable conclusion is still mind-boggling.

Through a further series of elegant thought experiments, Einstein showed that there is no notion of universal simultaneity; events considered simultaneous by one observer will occur at different times as far as another observer is concerned. The arrow of time, which had marched on at the same pace eternally, was free to speed up and slow down. Apparently there is no absolute concept of time, no universal clock out there in the sky to which we all must tune our watches. We keep our own times, march to our own rhythms. It is a wonder to me why the freedom to choose an individual perspective and claim it as valid should be so profoundly disturbing to mankind, but it was. It flew in the face of the rigid absolutes to which we were accustomed, and this unexpected liberation took some getting used to.

Since motion affects our perception of time, the same must also hold true for our notion of distance, if the speed of light is to be preserved. Following the same logic, Einstein argued that objects appear shorter when they are in motion (with respect to us) as opposed to when they are still. Does this not, then, distort reality, we wonder? No, says Einstein. The distortion is only in our perception. David sees Marcel's clocks as running slow, Marcel's rulers as having shrunk, but as far as Marcel is concerned, the world around him appears unchanged, and it is David's world that has been deformed.

Paradoxical as that may sound, it is no stranger than what occurs between you and your reflection in the hall of mirrors at an amusement park. Your image is bent and contorted to comic proportions—no sense of scale is preserved—and yet your reflection goes about life just as you do and is subject to the same laws of physics. One cannot but think that he finds your appearance quite as strange as you find his; which one of you, then, is correct? It is only fair to say: both.

Regardless of whether they agree with each other's perceptions, Marcel and David should be able to use their own observations to predict what the other will (claim to) see. But, as we already know, if these two inertial observers are to agree on the speed of light, their respective measurements cannot be connected by anything as straightforward as pure addition. The correct transformation rules are not difficult to work out; the problem is that they "mix up" space and time.

This sounds quite ridiculous. Space lies rolled out ahead of us; significant portions can be scanned in a single glance. Time can only be traversed sequentially; we must surrender one instant in order to move on to the next. Then, too, there is the obvious difference that we can move only ahead through time, whereas we circle around in space as a matter of course. Treating space and time as fundamentally different is quite understandable. Unfortunately, as Einstein found out, it is also quite incorrect.

Appearances can be deceiving, but mathematics does not lie. And if the equations say that space and time are linked, so it must be. I can understand easily enough that a line I see as horizontal will seem to have both horizontal and vertical components to an observer whose frame of reference is tilted. But the transformation laws of special relativity take this one step further when they say that, in a similar vein, a direction I see as space will appear to have both spatial and temporal components to an observer in a different inertial frame from my own. If they can be thus mixed up, space and time are clearly not so distinguishable as we had thought. In fact, the two should be combined into one larger entity, space-time, which has four dimensions (the three of conventional space, and one for time), so that transformations from one inertial frame to another may simply be thought of as rotations and translations in this new, four-dimensional arena.*

*It was Hermann Minkowski who first realized that Einstein's theory was best formulated in a four-dimensional space-time, where space and time "fade away into mere shadows, and only a kind of union of the two . . . [preserves] an independent reality."

This idea appears to be quite sensational when one first comes upon it, and much has been made of the "mysterious" fourth dimension. Yet, as Dr. Slosson reminds us, the laws of nature are not immutable or eternal edicts, but merely concise descriptions of how things behave. "There are no laws *in* Nature," he writes; "there are only laws *of* Nature; that is to say, laws drawn out of Nature . . . by man for his own convenience in thinking."

What we call laws are abstractions based on observation; they have not been handed down to us as gospel. Since our observations are limited, so, necessarily, is the scope of our laws. As we grow and learn, we often discover that these laws must be modified, and that is a change that a scientist should welcome. In fact, Dr. Slosson says, when a scientist is in possession of a theory that "will not hold half the facts he wants to put into it[,] he would have no more hesitation about dropping it than he has in setting down one beaker to pick up a larger one when what he has in the first is frothing over. He does not want to spill anything, but he does not care what vessel it is in."

Einstein showed that rods moving relative to us would appear contracted in the direction of motion and clocks moving relative to us would appear to run slow, but both effects would be practically negligible at the velocities we encounter on a regular basis. In other words, nature spares us the shock of facing distortions in our everyday mirrors; these absurd illusions are reserved for the "amusement park"—the realm we enter as we approach the speed of light. At the speed of light, time stops altogether and distances contract to nothing. One can go no further.

The speed of light sets the limit for all matter in the universe. Attaining this speed is like chasing the pot of gold at the end of the rainbow; no matter how far you go, the goal remains elusive. Only massless particles may ever achieve the speed of light; all material objects must resign themselves to moving more slowly. Nature enforces this decree in a rather clever manner: by making mass dependent on motion. The closer a massive object attempts to get to the speed of light, the heavier it becomes, until most of the energy channeled into making an object move faster contributes instead to making it heavier. Energy can do that, Einstein said, because matter is just energy made manifest.

In Battery Park yesterday, there was a sea of heads as far as the eye could see. Somewhere among the bobbing waves of gleaming copper locks, shiny top hats, and soft brown curls, my head of thinning black hair was bent over a beloved book. I felt the ripples of excitement flowing through the crowd before the shouts reached my ears. "He's here, he's here!" people cried out. Parents lifted children onto their shoulders to catch a glimpse of the great man. There was pushing and pulling all around me, but I dug my heels in and stubbornly held my ground.

And then, after hours of waiting, I saw a man in a faded gray raincoat, facing a veritable army of cameramen. His flopping black felt hat nearly concealed the gray hair straggling over his ears. Through the shouts and shrieks of the crowd, the incessant questions of the reporters, and the constant buzz of clicking cameras, he stood timidly, clutching a shiny briar pipe in one hand and a violin case in the other. The woman next to me turned to her husband in amazement. "He looks like an artist . . . or a musician," she said. I could not help the smile that floated to my lips. Little did she know, he was.

A chill went down my spine when I realized that Einstein was gazing in my direction. I felt my forehead wrinkle with intensity, as it always does when I concentrate. Please see me, please see me, I prayed. His eyes swept the crowd and, just for an instant, locked with mine. He smiled and nodded oh so slightly. All around me, people were moving, but I could have sworn I felt the march of time slow down.

When I came back to my surroundings, the cameras had stopped clicking. Einstein's motorcade had moved on, history had been witnessed, and the crowd was in a rush to disperse. The people in my vicinity glanced warily at me, probably wondering why I was standing still, but I was content to let the waves of the crowd break and wash over me. I did not want to break the spell of the moment.

At last I began winding my way home, taking care to avoid the main streets on the Lower East Side. I knew, as did the rest of the city, that Einstein was headed for the Hotel Commodore. Thousands of people lined the path, in a street party of sorts, waving to his motorcade. Even avoiding his route, I heard for a long time the honking of car horns, this modern form of fanfare being the tribute the streets of New York paid to the gentle genius. I wondered how he puts up with this constant rush of adoration around him. Perhaps he retreats to that secret "chest" in his mind, that imagined place, far out in dark, empty space, that served as a mental testing ground for the general theory of relativity.

Despite its phenomenal success, the special theory of relativity had an obvious limitation. By construction, it dealt only with objects that felt no external force, not even the gravitational pull of another massive body. For Einstein, this limitation was a problem. "No person whose mode of thought is logical can rest satisfied with this condition of things," he writes. In order to extend the validity of his theory, it was clear that he would need to incorporate gravity. But that involved first understanding what gravity is, so Einstein rounded up the facts as they existed at the time.

Beginning with the obvious, gravity is an attractive force that exists between any two massive objects. A defining characteristic of gravity, known as far back as Galileo, is that gravity causes all objects to accelerate at the same rate, regardless of their mass. In Newton's version of the story, gravity keeps us anchored to the ground, it pins the earth in orbit around the Sun, and it acts instantaneously over large distances.

Einstein found this tale to be somewhat one-dimensional. The existing narrative read as a list of events, of causes and effects. The predictions tallied well with reality, but the story felt flat. One could read it to the end and still not understand the protagonist. What is gravity really all about? What are its motivations for acting as it does? What are its demons, the realms into which it does not venture?

Einstein knew that, before he attempted a retelling of this classic tale, he had to ask some deep questions that would enable him to understand both the plot and the protagonist better. He started by focusing on the holes in Newton's narrative. First, there was the part about gravity acting mysteriously over large distances. This "action at a distance" had already been dealt with beautifully by Faraday, who invoked the concept of a field to explain a similar behavior in the electromagnetic force. Sensing the beauty and the merit of this construction, and feeling no need to reinvent the wheel, Einstein extended the mechanism to gravity. "Just as in electromagnetism, we think of a magnet attracting a piece of iron not directly but by calling into being something physically real in the space around it," Einstein wrote, we can think of two massive objects attracting each other across space through the medium of a gravitational field. Another problem was that, in Newton's description, the effects of gravity were felt instantaneously across space. But we knew this to be impossible, since nothing in the universe is allowed to break the (finite) speed limit imposed by light.

Einstein thought and thought and thought. For years he grappled with this problem, and then one day, in 1907, Einstein had what he later called the happiest thought of his life. As he sat in a chair in the patent office in Bern, all of a sudden it occurred to him that if a person falls freely he will not feel his own weight.* "This simple thought made a deep impression on me," Einstein wrote. "It impelled me toward a theory of gravitation."

Admittedly, "this simple thought" did not do quite the same to me. Thus far, I had managed to keep up with Einstein, but in that sentence he lost me. I backed up and thought it through,

---

*Free fall is defined as motion in the absence of all forces other than gravity.

slowly. A distant memory flashed into view. When I was growing up, a group of boys from my neighborhood would gather every evening in the empty lot at the corner and get up to the kind of antics that earned them the name "the hooligans." They didn't do anything really wrong, but they didn't seem to do anything very right either. I could never have been a part of them— my mother would have had a stroke—and besides, I wasn't exactly a daredevil and would not have fit in. But when I passed by the lot, I always slowed down a little and gazed out of the corner of my eye at this reckless world full of laughter.

Once, for a fleeting moment, I saw it from the inside. I was walking back home, my head hung down, and I had almost passed the lot when a voice pierced my thoughts. "Jacob. Hey, Jacob! Come here." I turned around, stunned, to find Isaac calling my name. Isaac was a classmate but not a friend, and I still do not know what possessed him to reach out to me that day. It turned out that, on that particular evening, the hooligans were attempting to fly. This involved scaling a broken wall, about eight or nine feet high, and jumping down. Isaac said it was quite the thrill and asked if I wanted to come try. I was scared out of my wits, but also extremely excited and somewhat flattered at having been included. Yes, I said, of course.

Somewhat in a blur, I climbed the wall with an army of boys clapping and hooting and cheering me on. Close your eyes on the way down, they shouted. Just close your eyes and jump. So I did. My entire flight lasted less than half a second (I worked that out from the relevant equations when I got home), but for that brief instant I really did feel free. Thinking back on the incident, I realized that the only time I felt a force was at the moment of my landing. The danger of the jump was not in the process but in its ending. I was reminded of Dr. Slosson's whimsical sentence: "The law of gravitation is like criminal law; you don't feel it till you come into conflict with it."

A feeling of comprehension slowly crept over me, and I thought quietly so as not to scare it away. While one is falling, one does not experience any force. Also, as Galileo proved, any two objects in free fall experience the same acceleration. Say, for example, that two boys jump down from the wall simultaneously. Since they both start from rest and experience the same acceleration, it follows that, at each instant during their fall, they will have the same velocity. This velocity will of course keep increasing, but since they keep pace so perfectly, they will appear—to each other—to be at rest!

Using this insight, it was quite simple to convince myself that two freely falling objects would see each other as either being at rest or undergoing "uniform motion in a straight line" (the latter case being when, for example, one object has been in free fall longer than the second and thus has a head start on the velocity).

Although intellectually a little winded, I finally caught up to the page where Einstein sat patiently, waiting for me. He continued: It is settled that when you fall in a gravitational field, you feel no force, and that an object falling with you will appear to be at rest. Gravity, as we have already established, is a variable force, its strength adjusting automatically so that it causes every massive body to accelerate at the predetermined rate. This fact suggests a deep connection between the nature of gravity and acceleration, and Einstein was determined to hunt it down.

It was then that he invented that chest in space, the one that was large enough to be a room, but with no peepholes to the outside. David lives inside this chest, whereas Marcel stands and watches from a fair distance. As long as the chest is far away from any gravitational field, David is an inertial observer who goes about his business unfettered, as inertial observers do. Since he is not subject to any forces, David is free to float in midair and stay perfectly still—there is nothing to pull him down. If, however, *he* exerts a force—say, by pushing against a wall— the corresponding (equal and opposite) reaction will impel him into motion. He will drift around the room, perhaps colliding with other floating objects, any such impact modifying their motion just as if they were billiard balls on a table.

ONLY THE LONGEST THREADS

This scenario seems to make sense, so we take it one step further. Suppose now that the entire chest is accelerated upward at a constant rate. If David stands on the base of the chest, the floor pushes up against him as it rises, and in reaction, his feet push down on the floor. Consequently, David experiences the familiar sensation of weight. But what happens to an object that is not in contact with the chest? Say David, who has become accustomed to living in the absence of forces, places an apple in the air. He expects it to stay perfectly still until he reaches out for it again, but now that the chest is accelerating, this is not what unfolds. The apple falls to the floor, with a constant acceleration; this behavior, so reminiscent of his earthly experiences, leads David to conclude that he must be at rest in a gravitational field. Marcel is able to see that there is no gravitational field; that the apple does in fact stay suspended in space, and it hits the floor only because the chest on its accelerated path upward "rises" to meet the apple.

Faced, once again, with two equally valid perspectives on the same situation, Einstein began to wonder if the two descriptions were in fact the same. Perhaps the effects of (uniform) acceleration are truly indistinguishable from those of gravity. To test this idea, he applied it to another accelerating system. Objects undergoing rotation provide wonderful illustrations of accelerating systems, and luckily we have an obvious example at our disposal: the Earth.

Newton formulated his laws of motion treating the Earth as stationary—i.e., in a frame of reference that rotates along with the earth—but this motion would be evident to an observer who peered at the Earth from a static frame of reference, situated far off in space. To such an observer, the equatorial bulge would seem a natural consequence of the Earth's rotation, but those of us who move with the Earth (and hence are numb to its motion) attribute the same bulge to a centrifugal force. The centrifugal force is so similar in character and effect to gravity that we do not even bother to distinguish them. Instead, we absorb the effects of the centrifugal force into the earth's gravitational field, and treat the latter as if it varied across latitudes. Since a force that has obvious, visible implications (like the equatorial bulge) can be made to appear and disappear merely by changing frames of reference, we must conclude that forces too are relative.

Working through a few more thought experiments, Einstein eventually convinced himself that, just like inertial observers, accelerating observers too can claim to be at rest—provided they pin the blame for their observations on a suitable gravitational field. Before we make any measurements, we have to choose a measuring system, or a frame of reference. Measurements are always made against a set of standards, and we have to pick ours. This choice influences what we see: it determines whether or not we will perceive a force or sense a magnetic field, and it affects the numerical values of the measurements we make. But it is important to realize that our need to choose a perspective is a practical concern, not a constraint on nature. Physics prefers no one frame over another; the laws of nature should apply equally everywhere.

When I reached this stage in Einstein's book, I had a headache bordering on vertigo. These ideas were dizzying in their intensity, and I could feel that the crescendo was close. Part of the problem with Einstein is that he does not ask questions one can walk away from. He will not say, as we expect scientists to: "What happens if I change this equation around and add a term here?" Instead he flashes an irresistibly vivid image. "What would I see if I rode light?" he asks, or "What would I feel if I fell off a roof?" His way of phrasing problems pulls you in, and before you know it, you are standing at the mouth of a labyrinth.

It is hard to argue with the steps Einstein asks you to take, as he leads you patiently past obstacles you could never have navigated alone, until without warning you find yourself out in the sun again, but in a world quite different from the one you left behind. There *is* a logic to the process, but there is also a sense that the process was not discovered by logic alone.

Perhaps Einstein really does have "a speculative imagination so vast that it senses great natural laws long before the reasoning faculty grasps and defines them." People have remarked that this uncanny ability to visualize scenarios so vividly might, in part, be due to his thorough examination of patent applications. His facility for conjuring up imagined machinery must have been strengthened in the hours spent poring over technical designs, checking to see whether or not they would work.

When the distinguished guests arrived at Battery Park yesterday, reporters were, of course, asking questions of everyone in the party. To Chaim Weizmann (Einstein's friend and traveling companion) they addressed the inevitable good-humored question: Did he understand his friend's theory? Weizmann replied, "Einstein explained his theory to me every day, and on my arrival I was fully convinced he understood it." The crowd broke into laughter, but I kept chuckling after the others had stopped. It was reassuring to hear that even the privileged few who were granted such intimate access to Einstein did not find the theory simple to understand. With that one sentence, my sleepless nights and throbbing head were vindicated.

But coming back now to gravity. If it is indeed interchangeable with acceleration, Einstein said, we can gain insights into the behavior of objects in a gravitational field by studying how they are affected by acceleration. With this aim in mind, we return once again to that windowless room out in space, roll up our sleeves, and get to work. Suppose David shoots a ray of light from one end of the room to the other. As light covers this distance, the room accelerates upward. By the time the beam arrives at the opposite wall, the floor has risen from its original position, so the point at which light hits the wall is considerably lower than the point from where it set out. Moreover, as the astonished David realizes, the path taken by the light is not just tilted downward, it is curved. This result was exceedingly strange. Light is supposed to travel in straight lines, but somehow acceleration plays havoc with this long-accepted fact; lines are bent into curves, geometry is modified.

It is not that parabolic motion in itself is surprising. Falling objects have long been known to trace curved paths in their descent to the ground, and the equations describing this projectile motion have been used for centuries. The problem was that such behavior was expected only from masses, upon which gravity exerts an attraction. Light, being massless, was supposed to be exempt from participating in this interaction.

But, since Einstein had convinced himself that light would appear to be bent in an accelerating frame, he had to find a way to explain this in a gravitational context also. Such an explanation is difficult to come by if we continue to think of gravity in the manner of Newton, as a mysterious influence tying two material objects together. What if we shift perspective?

There had always been something subtly different about gravity. Though we call it a force, it stood a little apart from others. Dr. Slosson terms it "unique, independent, irreducible, unalterable and inexplicable." Under the thrall of the other forces, he writes, different substances behave differently: "one is more easily heated than another; some are readily magnetized or electrified, others are not so susceptible; certain elements rush into each others' arms, others cannot be forced into combination. But gravitation seemed indifferent to all these things; it showed no prejudices or preferences. It attracted with equal force all sorts of substances, no matter whether they were hot or cold, shiny or black, moving or still, electrified or magnetized or neither. . . . All other forces could be reduced or increased, annulled or brought into effect at will. Not so gravitation. Any bodies of a certain mass placed at a certain distance apart are always drawn by the same attraction. That is, gravitation is affected by nothing except geometrical relationships."

What if this behavior hints at a deeper truth? Einstein asked. What if gravity *is* in fact a manifestation of the geometry of space-time? With this new perspective, everything snaps

ONLY THE LONGEST THREADS

into place, as long as we allow for this geometry to be curved. A precise formulation of this statement requires stronger command of mathematics than mine, but the general idea makes sense. The rules of geometry are different for spheres than they are for sheets. We tend to call upon Euclid's axioms unthinkingly, but these tenets hold only for flat surfaces. If circles with equal radii are drawn on the inside of a large round bowl, a flat sheet of paper, and the surface of a ball, the circumference of the first will be larger than that of the second, and that of the third will be smaller still. Other familiar truths change as well; for instance, the sum of the interior angles of a triangle is 180 degrees only if the triangle is drawn on a flat surface. But perhaps most surprising of all is the discovery that a straight line is not necessarily the shortest distance between two points.

To see why, mark off two points on a circle: call them A and B. Starting from A, there are two distinct ways of getting to B (moving clockwise or counterclockwise), and one way will be shorter than the other.* Notice, however, that both these paths are curved. As long as we are constrained to move only on the circle, a straight line connecting A and B (in other words, a chord) is no longer an option. The shortest possible path is curved, simply because that is the nature of the space. Newton's law of inertia—with its insistence on "uniform motion in a straight line"—was clearly formulated for flat Euclidean space, but it can be extended quite naturally to curved space if we decree that inertial objects travel along the shortest paths they can find. In flat space, these paths happen to be straight lines; the curved space analogues are called geodesics.

Einstein's calculations showed that gravity changes the shape of space-time, just as our study of accelerating systems had hinted. Every massive object exerts a gravitational force, and thus every mass changes the shape of space-time. Heavier masses

---

*Unless A and B are diametrically opposed, in which case both routes will have exactly the same length.

change it more and lighter masses change it less, but each mass makes its presence felt.

An obvious problem with talking about "the shape of space-time" is the fact that space-time is intangible. So what exactly is it that curves? Without delving into higher mathematics, all we can say is that the arena in which we find ourselves, our entire playing field, changes shape. It is a rough analogy, of course, but we are sketching with a few words what can be exactly stated only in pages of mathematical equations.

It remains now only to tie up the loose ends. We have concluded that every massive object changes the space around it, and that this effect is strongest close to the object. Consequently, inertial observers find themselves tracing curved paths in the vicinity of a mass. Dr. Slosson describes this rather eloquently: Imagine a sheet of rubber stretched over a hoop until it is taut like a drumhead. Now draw a grid on this sheet, with each square of equal size. Two or more worms, crawling on parallel lines, will stay equidistant from each other throughout their entire journey. If, however, a massive object (say a bullet) is laid in the center, the rubber sheet will sag and stretch. The squares we drew on the rubber will now be deformed so that they no longer have equal areas; lines that were parallel cease to be so. This effect will be most extreme in the center and get progressively less pronounced toward the edges. If our worms traverse this terrain again, they will find their experience much changed. Those who follow lines drawn close to the bullet will need to dip down into a valley and climb up again, thus traveling a greater distance, and consequently taking longer, than the worms who crawl along the (nearly unchanged) lines lying close to the edge.

Having witnessed this behavior, we could explain it in one of two ways: we could assume that "the worms on seeing the bullet to one side were drawn by their curiosity a little toward it, those nearest of course being drawn the most," writes Slosson, or "we might assume the existence of a 'force' in the bullet which in some mysterious manner attracts the heads of the worms inversely as the square of their distance." But instead of attempting to delve into the psychology of worms, or postulating an invisible cord or an incomprehensible force, "is it not simpler to consider the space between and to suppose that the lines to be traversed are lengthened in the neighborhood of the weight?"

What we call gravity, Einstein said, is merely a manifestation of this warped geometry. If this is indeed true, if the very shape of space-time itself changes, then it is not just massive objects whose paths are affected; light, too, will be forced to bend according to the dips and curves it encounters in its journey from one point to another. This is where Einstein's famous result came in, the one that erupted in headlines across the globe. If the theory is true, the light from distant stars should change its path a little when it grazes the sun. Einstein's equations even predicted the precise degree to which light should turn. When this effect was finally measured, in the solar eclipse of 1919, he was vindicated.

Late last night, so caught up in my thoughts that I couldn't sleep, I stepped outside to clear my head. "Hey, Jacob," I heard someone call. My neighbor on Henry Street, Tony, was sitting on his stoop, smoking a cigarette. He told me that his cousin had just come through Grand Central, and as of 11:00 p.m., Einstein's cavalcade had still not reached the Hotel Commodore! What a reception the people gave him, Tony muttered, what a guy! I nodded agreement.

Now that I had snapped out of my reverie, my aching limbs reminded me how tired I was. I told Tony I would catch up with him in the morning and climbed the stairs to my apartment. I opened my little window a crack, to let the fresh air in, and found myself gazing at the night sky as if I had never seen it before. Who would have suspected that starlight does not whiz nonchalantly by the sun but in fact tips its head in acknowledgment? In the dark depths of space, completely unbeknownst to us, these subtle social graces have been exchanged for centuries. How many other secrets, I wondered, do the heavens hide? Which of nature's refined mannerisms are we still missing out on?

CHAPTER 4

A Movie Comedian or Something

[Quantum Mechanics]

Physics is very muddled again at the moment;
it is much too hard for me anyway, and I wish I were a
movie comedian or something like that and had never
heard anything about physics! —WOLFGANG PAULI

APRIL 1930
BLEGDAMSVEJ 17, COPENHAGEN

My dearest Anna,

I write these lines sitting in what I have come to think of as our place, in the middle of the third row. If you were here, I like to think this is where we would come in the evenings to talk. I guard my excursions to Auditorium A as jealously as if they really were assignations, even though for now the only tryst is between my thoughts and me. Innocent as the activity is, my heart does beat a bit faster when I walk here, ready to modify my route at a moment's notice should I encounter people in the corridors. I always have some excuse ready to hand out to those I meet; I am going for a walk, or a drink of water . . .

I close the door behind me, because I would just as soon avoid explaining why I am sitting here, alone, rather than in the library or an office. I doubt they would think it odd, for no one in these parts is a stranger to eccentricity, but they are—one and all—given to constant debate and discussion; to enter even a casual conversation is to become embroiled in a quantum tangle. And while I love trying to wriggle my way out of those tight knots, my arguments bending over and around themselves as if they were contortionists on stage at the Tivoli Gardens, this is my time with you.

I particularly love this room, because I feel the palpable presence of ideas hanging thick in the air. Countless concepts have collided here, some ricocheting off others, some fusing together, and others being cracked open; the sounds of those impacts echo still. Embossed into the blackboard are layers of equations, written and rewritten, erased, corrected, forgotten, enshrined. Looking down from the walls are the black-and-white faces of the insightful few who have, in these past few years, peered into the atom. In a room such as this, how can one help but feel inspired? If one must come to grips with quantum theory, what better place can there possibly be than this room where—more than anywhere else—it lives?

Perhaps that last sentence might sound strange to some, but you who so often mirror my thoughts will know that it is not quite strange enough. No matter how much we try, quantum theory refuses to be pinned down. It will not give definite answers to our questions. There are no certainties anymore, only probabilities, so it is only fitting that no single place be declared the birthplace of quantum theory. The best one can do is to say that the theory was born at various times in various places all over the globe. But still, if I had to calculate the coordinates of the place quantum mechanics calls home, I suspect that the wave function would be strongest here in Auditorium A at Blegdamsvej 17.

Over the years, almost everyone who has ever worked on quantum mechanics has passed through these doors in Copenhagen, leaving their chalk-dust footprints behind. The theory wanders these halls, spreading like a wave; it ripples through windows, diffracts through keyholes, and tunnels through walls. At other times, it shoots out idea quanta in focused beams, pelting them at every conceivable surface, until the rays rebound, reflect, and finally pass into notebooks or blackboards, refracting slightly as they do so.

When Bohr opened these doors to quantum mechanics, it was just a fledgling theory—iconoclastic, rebellious, and awkward like a gawky teenager. Most traditional departments looked upon it cautiously as if it were something to be handled with tongs. But, like all adolescents, quantum theory just wanted to feel welcome; it needed to be surrounded by young people who were not yet completely indoctrinated by classical mechanics and were still open to wild new ideas. Bohr's Institute of Theoretical Physics, with its extremely informal atmosphere, was exactly the sort of foster home this juvenile theory needed to thrive.

Bohr's own work appears to be shaped almost entirely in conversation. He thinks about problems so obsessively that eventually he begins to sense relationships intuitively, without having to derive them formally. Through gentle probing questions, ceaseless encouragement, and examination, he coaxes out ideas from his mind and others', relentlessly smoking a pipe the entire time. His fledgling institute grew around these mannerisms; it became a place where debate was incessant, but pomposity frowned upon. Individuality and creativity were prized, yet a friendly collegiality inevitably emerged from among the chaos and chatter.

In some ways, Blegdamsvej 17 reminds me of the Never-Never Land I so longed to find as a child. I can easily picture Bohr as Peter Pan, leader of a troop of brilliant young lads in their twenties who treat quantum mechanics as an elaborate game to play, in between ping-pong on the library tables, cowboy movies, and mock shoot-outs with toy pistols. Like children playing make-believe, they merrily bend and break the rules of "reality," putting up a brave fight against every fresh observation that experiments throw at them, fending them off as if they were pirate attacks. Ideas are tossed back and forth in endless sport. No one sits on the sidelines for long; everyone in the crowd is also a player, and the energy is contagious. The game never stops. Almost insolently, just as in the storybook, these Lost Boys disdain age. In Blegdamsvej folklore, "when one is past thirty, he is as good as dead." Quantum mechanics truly is a playground for the young. And that is sometimes a frightening thought for one who, like me, stands on the cusp.

The Institute seems ordinary enough from the outside; there is nothing to distinguish it from its neighbors, nothing to warn a passerby of the madness that reigns within these walls. Often I think we should follow the custom of Copenhagen's shopkeepers and hang up a metal sign above the front door; but what symbol would we employ? The pet store displays the silhouette of a parrot, grapes announce the liquor store, a pastry dangles above the bakery, and a watering can and kettle are suspended above the metalworker's shop: each simple image announces its occupant's vocation and business with clarity. Part of the trouble with quantum mechanics is that we don't have such a symbol yet. The world that exists this side of the threshold is surreal, but it has a strange logic all its own—unexpected, but self-consistent.

Drawn by Bohr's work as well as by his humanity, visitors flock to Copenhagen. Last year Bohr instituted an annual conference as well. At his invitation, the world's most brilliant minds come here in the spring to assemble quantum theory piece by piece. People are everywhere; they fill the rooms and spill into the corridors. Everywhere, passionate conversations are to be heard. The pace is frantic but discussions are organic, evolving spontaneously according to the needs of the moment; there is no set agenda. In keeping with the casual mood of the Institute, the traditional end-of-the-meeting summary is presented as a parody. It is a good opportunity to poke fun at ourselves, each other, and the work that consumes our days and our minds. This room where I now sit becomes a beehive of activity. At the blackboard before me, excitable young men fight out intense duels, waving long pieces of chalk like rapiers, defending theories as medieval knights defended their honor.

I look up to my right, at the wall where the conference is frozen in photographs. Each distinct, intense face is etched in my mind from the many times I have stared at these pictures and the many times I have seen the people themselves, in all their animated zeal. But right now, in the fading light and from a distance, all I can make out is a haze of black and white. From somewhere in this mist where distinct features are lost and boundaries blur, a faint hum emanates, and it strikes me afresh that no single man, no matter how talented, could have sustained all these distinct voices in his head without going insane.

Often people comment on how different the collegial Copenhagen meetings are from the formal Solvay Conferences, and yet what strikes me most is that they both germinated from the same seed: the recognition that, for the first time ever, physics had to be done by committee. In a way, the existence of these conferences is an open admission that what faces the world right now is a madness too large to be contained in a single mind.

I settle deeper into my seat. It really is the perfect spot. We have chosen our place well. Shafts of afternoon sunshine illuminate the pages of my notebook, infusing them with a dull glow. Such a gentle, reassuring guise it is that light assumes, and yet how deceptive is that impression! Light has led us along many a tangled path for centuries past; from Newton to Maxwell to Einstein, it has tantalized the minds of many great physicists, illuminating further ground, marking a trail leading them to strange realms of thought unheard of before. Perhaps Einstein's axiom was true in more ways than even he suspected—perhaps we really cannot catch up with light, no matter how fast we run!

Not content with revolutionizing electrodynamics and forcing the birth of relativity, light lured physicists into another maddening quandary. At the turn of this century, the "peacefully inclined" and unadventurous Max Planck reluctantly unleashed the biggest revolution physics had known since the days of Newton. While struggling with the stubbornly intractable problem of the absorption and radiation of light by material bodies, Planck realized that all the paradoxes that had been plaguing our understanding for years would disappear if we assumed that energy could be radiated and absorbed only in discrete blocks, which he called light quanta. Upon further investigation, it was discovered that the size of this energy block was directly proportional to the frequency of the emitted radiation.

Radiation was presumed to arise from atomic oscillations, and for a while Planck held on to the hope that the presence of quanta was a reflection of the structure of the atoms, not a phenomenon intrinsic to light. It is easy to sympathize with Planck's desperate avoidance of the consequences of his own declarations, because, if you think about it, the notion of a quantum changes everything! Quanta carried with them the explicit statement of discontinuity, in a world where most existing theories were formulated on the assumption of the infinitesimal: the belief that no matter how small a quantity, it could always be divided further. In the quantum realm, gradual transitions are no longer the norm; this is not a place in which you emerge gradually from the shadows into the light—you are abruptly and unceremoniously thrown out. The seasons do not seamlessly dance into each other; night no longer melts into day; there is no infinity of in-betweens.

The familiar sweeping continuum of nature was replaced by a grid. Being and nothingness were separated by a sharp line, and Zeno's paradox became an oxymoron. Planck gazed, horrified, into the void, but there was no going back. The age of the quantum had arrived. Once they got over the initial shock, people began to explore whether this idea could be used to make sense of other inexplicable processes. One such phenomenon waiting for an explanation was the photoelectric effect.

Around the same time that Planck was cornered by the quantum, atoms were found to be undeserving of the epithet bestowed upon them by the Greeks, as experiments revealed the composite nature of these supposedly fundamental entities. Though electrically neutral as a whole, atoms were found to contain equal amounts of positive and negative charge. The negative charge was distributed equally among particles called electrons that ran circles around the nucleus, the seat of all the positive charge. Inevitably, a comparison was made with the heavens; the image of an atom as a miniature solar system held undeniable philosophical appeal.

The nucleus, the would-be sun in this system, was disproportionately small, given the size of the atom. In fact, Rutherford, who "saw" it first, said it was like "a gnat in Albert Hall"!

Matter, which we instinctively thought of as being uniformly solid, thickly packed and substantial, was exposed as being porous. The emptiness within the atom eclipsed the emptiness of interstellar space as disconcerting truths came to light. It was estimated that if all the empty space within atoms were eliminated, and the nuclei and electrons packed tightly together, a man would shrivel to a speck, barely visible with a magnifying glass. I think of the grandiose claims of the Ancients, who stated with perfect confidence that nothing existed except atoms and empty space. I wonder how they would have reacted if they were told that empty space encroaches upon atoms too, and as a result renders us all, you and I and Greek philosophers alike, surprisingly vacuous.

A further conundrum was raised by the photoelectric effect, whereby electrons are emitted from a metallic surface upon which light is shone. The phenomenon in itself was not surprising. As the image of the atom evolved, it became an accepted fact that electrons are held in an atom by their electrostatic attraction to the nucleus. It made perfect sense for impinging light beams to supply electrons with energy, enabling them to break their shackles and escape. The inexplicable part of this tale was the fact that the intensity of the beam did not seem to affect the energy of the electrons emitted; it influenced only their number.

What was even stranger was that, regardless of the intensity of the beam, no electrons were emitted at all unless the frequency of the light source was above a certain threshold. Had light consisted of waves, as classical theories suggested, these waves would keep washing over the electron, building up slowly, until a large enough tide was created to carry the electron right out of the atom. But experiments showed that this was not the case. If, for some reason, the electrons had resolved to wait for blue light, no amount of red light showered upon them would make them give in; they were passionate in their demand for, and response to, the proper color. Why would this be so?

Once again it was Einstein, with his uncanny insight into the affairs of light, who answered this question. To Planck's growing unease, Einstein conjectured that light itself is inherently grainy. He pictured a beam of light as a stream of indivisible particle-like entities called photons, each carrying a unit of energy whose size is determined by the frequency (or color) of the light. An intense beam has *more* photons than a weaker beam of the same frequency, but individual photons in each beam carry the same amount of energy. While it is still true that electrons depend upon photons to help them escape from the atom, the necessary energy must come from a single photon—it cannot be built up bit by bit. The barrier erected by the atom, Einstein said, must always be surmounted in a single jump.

If the amount of energy contained in an individual photon is tied up with the frequency of the beam, it is easy to see why no electrons are emitted in response to an intense beam of low frequency; no matter how many of these photons are hurled at the metal, none of them carry the quantum of energy an electron needs to break its bonds. When the frequency of light is increased, the amount of energy carried by each photon goes up. As soon as this quantum of energy is sufficient to do the deed, an electron absorbs a photon, thereby gaining the strength to jump the barrier erected by the atom and roam free. Moreover, Einstein said, if the energy of the photon is greater than the binding energy, the electron will carry away the "excess" energy as kinetic energy. This last fact was checked experimentally and found to hold true, confirming Einstein's theory.

If any further proof of the particulate nature of light was needed, it was provided by Arthur Compton, who showed in a brilliant experiment that when X-rays are scattered off electrons, their interaction mimics the collision between two rubber balls. So, in the space of one eventful decade, it was uncovered that light was a stream of tiny particles and the atom was a miniature solar system. All these modifications in our system of thought were difficult even for the experts to take. According to Eddington, James Jeans once said, "not only does the quantum theory forbid us to kill two birds with one stone; it will not even let us kill one bird with two stones!"

There was no question about it. The quantum challenged all our previous assumptions, and yet this outrageous idea was borne out by experiment. The chief obstacle that had to be overcome was the desire to understand a phenomenon by drawing parallels with everyday life. As Sir Arthur Eddington reminds us, even though science aims to construct a symbolic world that represents reality, "it is not at all necessary that every individual symbol that is used should represent something in common experience or even something explicable in terms of common experience." In fact, such efforts are doomed to fail. In order to convince us to relax the need for literal interpretation, Eddington points to a familiar operation: reading. The intention is to communicate with the reader by having him associate inky marks on paper with concepts and experiences drawn from life, but the individual letters in themselves do not have any natural counterparts in reality. These fundamental symbols are abstractions and must be understood as such. Just as A stands for itself, not an Archer or an Apple pie, says Eddington, we must come to accept that the only way to understand an electron is to say that "it is part of the A B C of physics." In order to explore the nature of this new alphabet of nature, we must remove the assumptions we have unthinkingly forced upon the letters, causing them to squeeze and deform until they bear no resemblance to their proper shape.

These shifting foundations of reality caused the world's most spectacular and renowned physicists to convene at the first Solvay Conference. With brilliance, diligence, and remarkable open-mindedness, the exclusive list of invitees set upon the task of deciding whether or not the quantum was to be included in the physicist's repertoire. Did Nature paint with the flowing brush-strokes of the old masters, they asked, or with the tiny dots of Seurat? The elite jury members were enjoined to "take so fruitful a theory seriously and subject it to careful investigations" before finally pronouncing their judgment.

Smoldering with purpose and passion, they carried the torch back to their own institutions, setting colleagues ablaze. Wordlessly, the physics community united to forge some sense out of the quantum. Niels Bohr, then a student at Manchester, caught the fire directly from his professor, Ernest Rutherford, a Nobel laureate and one of the handpicked attendees. Soft-spoken Bohr with his vaguely embarrassed smile might have seemed an unlikely champion for the rebellious and unruly quantum theory, but he was one of its first.

Even if one accepted the dissection of nature into tiny pieces, there were several troubling issues still to resolve. One such problem had to do with the origin of atomic spectra. It had long been known that when a pure element is heated until it glows, and the resulting light passed through a prism, we see a series of distinct bands separated by empty space. These patterns—or spectral lines—are unique to each element and can thus be used for identification purposes. For years, they were regarded as an artistic flourish of nature, much like the "lovely patterns on the wings of butterflies," but with all the new discoveries surrounding the atom, people started questioning whether these atomic signatures pointed to a deeper truth. Why would the radiation emitted by a particular element contain only a certain set of frequencies? Did the spectral lines somehow encode information about the structure of the atom?

The planetary model of the atom, which had been authored by Rutherford, was not problem free either. Maxwell's electrodynamics said that an electron in orbit, like any accelerating charged particle in motion, should radiate energy continuously. If this were indeed the case, electrons in atoms would constantly lose energy, slowing down and spiraling inward until, with too little energy left to sustain motion, they would succumb to the pull of the nucleus. According to calculations, this atomic collapse should occur within fractions of a millisecond. Experience, of course, tells us that this is not so; atoms are extremely stable, and it was obvious from our faulty conclusions that an understanding of the relevant physics was still lacking.

In 1913, with the cavalier confidence of youth, Bohr proposed a possible resolution. The contradictions could be eliminated, he said, if we liberate the atoms from obeying Maxwell's laws. We have no prior experience of objects this tiny, he said. Our intuition and expectations are shaped by what we perceive around us, at scales that are roughly a thousand billion times larger than atoms. The very existence of a quantum is an indication that life on the subatomic level is vastly different from what we are used to. Perhaps a different set of rules applies in that domain.

Bohr pictured the vast expanse of the atom as containing a finite number of allowed orbits into which electrons could slip; each orbit had an associated energy, labeled by an integer called a quantum number. While an electron was sliding around in one of these invisible grooves, or "stationary states" as Bohr called them, it would maintain the corresponding constant energy. Only when an electron jumped from one orbit to another would it emit radiation, the quantity of which was equal to the difference in energy between the initial and final states and had nothing at all to do with the orbital motion of the electron. We know from the sharp, distinct lines in atomic spectra that this emitted radiation is quantized, and thus it stands to reason that the mechanical energies of the electrons in orbit must also be similarly constrained. Further, said Bohr, what distinguishes one element from another is the fact that their atoms have different allowed energies. Using this model, Bohr was able to calculate, quite accurately, the well-known emission spectrum of hydrogen as well as the spectra of heavier elements.

More sensitive experiments revealed discrepancies between predictions and observations. In an uncanny instance of history repeating itself, Arnold Sommerfeld came to the rescue, playing Kepler to Bohr's Copernicus. Centuries earlier, Kepler had found that planets in our solar system did not orbit the sun in perfect circles, as Copernicus had suggested, but instead in ellipses. Sommerfeld made exactly the same suggestion for electrons revolving around a nucleus. Practically speaking, this meant that an orbit could no longer be specified by a single quantum number alone; three parameters were needed. It is easy to see why. In order to describe a circle (in a fixed plane, and with a fixed center), all we need to know is the radius; but an ellipse is characterized by two numbers—the lengths of its major and minor axes. Moreover, to properly situate an ellipse in three-dimensional space, we need to specify the plane in which it lies. In keeping with the traditions of quantum mechanics,* all three of these numbers were integers. The theory, thus modified,† worked so beautifully that Sommerfeld was moved to compare the spectra to "a veritable atomic music of the spheres."

In 1922, Bohr, who was firmly established as one of the pioneers of quantum mechanics, was invited to give a (hugely successful) series of lectures on atomic theory at Göttingen. In the wake of this triumph, his popularity soared even further, and

*The fact that a quantum exists means that every quantity can be expressed as an integer multiple of this smallest basic unit. Hence, to specify an amount, all we need to know is the number of quanta present, i.e., the quantum number. Needless to say, this is always an integer.

†Sommerfeld also treated the hydrogen atom relativistically, because the mass of the electron increases with its velocity.

many talented young scientists were attracted to Copenhagen, where Bohr supported them with funds provided by the Carlsberg brewery. Is it any wonder, then, that this new science is so effervescent and giddy?

The precocious duo of Wolfgang Pauli and Werner Heisenberg was instantly drawn to this hive of activity. Pauli fast became Bohr's favorite sparring partner, and Heisenberg a cherished friend, but the modus operandi of his two young disciples differed widely from Bohr's primarily philosophical approach. Even though neither of the two is resident here anymore, both are frequent—and very welcome—guests at the Institute, being brilliant physicists and much loved by Bohr.

Pauli is somewhat of an acquired taste. He is caustic, witty, and extremely passionate; when he defends an idea, his whole corpulent body oscillates with excitement. In our small community of quantum mechanics, ideas are not considered developed until they have passed Pauli's test. He has an uncanny ability to see the holes in any theory, no matter how well they are disguised. Even though he is much admired, he is also faintly feared. Some of the younger students hold that Pauli will mysteriously appear, from amid a fog of sardonic laughter, if you speak of theoretical physics for too long. The number of myths that surround him already is astounding, but none is stranger than the so-called Pauli effect.

It seems to be accepted almost as fact that theoretical physicists break experimental equipment whenever they handle it, but Pauli takes this to a whole new level. The claim is that Pauli has such a good command of theory, he breaks things just by crossing the threshold of a lab. A particularly entertaining anecdote concerns a mysterious experimental collapse in Göttingen. People at that laboratory wrote humorously about the incident to Pauli, saying this was the kind of thing they would expect his presence to cause. They mailed the letter to his Zurich address but received a reply in an envelope bearing a Danish postmark.

It turned out that Pauli had been on his way to Copenhagen, and at the time of the accident his train had stopped over at Göttingen!

Pauli was first drawn to quantum theory by Sommerfeld's lectures in Munich. The Bohr-Sommerfeld model of the atom intrigued him. Despite its success, this model was essentially just an empirical formula that fit a set of observations. It had predictive power, but since it had not been derived from any deeper principles, it lacked the ability to explain why things were so. Still, using this model, one could calculate the values of the discrete atomic energies assigned to each element. These energies fell into a neat, aesthetically appealing pattern, which was intellectually satisfying to many, but even as a student Pauli was not seduced by what he called this "number mysticism" everyone else advocated. He wanted to know why nature chose the particular values it did, passing over all other possibilities.

Further investigations revealed that the Bohr-Sommerfeld model explained, almost perfectly, the behavior of all the elements in the periodic table, as long as there were *twice* as many electrons in each orbit. This doubling meant that a fourth quantum number was now needed, to distinguish between two electrons for whom the existing three quantum numbers (describing the orbit) were identical. This new quantum number need only have two values,* but physicists were perplexed as to how it should be interpreted. What physical property could it possibly correspond to?

Faced with this conundrum, the usually cautious Pauli proposed a rather radical solution: what if, he said, this fourth quantum number referred not to the orbit in which the electron revolved, but instead to an intrinsic attribute of the electron itself? If so, the electron would need to be possessed of a property with

*Only two distinct labels are needed to label two objects in a unique manner.

respect to which it could take two values. This "two-valuedness" was strange, not only because it had no classical counterpart, but also because consistency of the theory required these values to be half integers! Whatever the physical quality corresponding to this quantum number, it was not a binary property that the electron either carried or didn't; the two allowed values for the quantum number were not 0 and 1. The quantum number represented something more akin to charge, in that there was a positive and negative value, but these values could not be stated as +1 and −1—they insisted on being +½ and −½.

To add insult to injury, Pauli, who had objected to the lack of a derivation for the previous three quantum numbers, could do no better for this fourth. In fact, it was worse—he could not offer a physical interpretation either! All he could do was argue that this suggestion appeared to "offer itself automatically in a natural way." This sudden introduction of half integers by the theory's most conservative and critical practitioner, the man who disdained "atomysticism," was thought to be hilarious by many who had weathered his scathing wit.

Heisenberg, in particular, was especially amused. A few years earlier, when Heisenberg had attempted to explain a puzzling feature* of the atomic model by suggesting that the quantum numbers *might* be able to take on half-integer values, Pauli had ruthlessly criticized the proposal in his characteristic manner, saying, "Now you introduce half quantum numbers, then you will introduce quarters and eighths as well until finally the whole quantum theory will crumble to dust in your capable hands." When a twist of fate had Pauli postulating an entirely new quantum number, which could take on *only* half-integer values, Heisenberg could not contain his glee. He wrote to Pauli saying how pleased he was that "you have raised the swindle to unimagined, dizzying heights . . . [breaking] all previous records you accused me of holding."

---

*The anomalous Zeeman effect.

The teasing aside, it appeared that Pauli had stumbled on something fundamental. His so-called Exclusion Principle stated that every electron in an atom must occupy a unique state, and hence must carry a distinct set of quantum numbers. The distribution of electrons in an atom—or, indeed, any system—can then be attributed to the fact that each electron tends to minimize its energy by occupying the lowest possible energy state. But, even though the Exclusion Principle worked so well, the problem of interpreting the new quantum number remained, until Samuel Goudsmit and George Uhlenbeck put forth the concept of spin. Drawing yet another analogy with the solar system, these two young students of Paul Ehrenfest conjectured that an electron spun on its own axis, much like the earth. The two values ($+\frac{1}{2}$ and $-\frac{1}{2}$) of the new quantum number, they said, corresponded to the electron's rotating clockwise or counterclockwise.

Their proposal met with a remarkable amount of success, and Pauli is said to have regretted not putting this explanation forward himself. It had occurred to his assistant Ralph Kronig too, but Pauli had brushed it away. A true quantum thinker, Pauli wants to emit theories only when they are fully formed and can be phrased in terms of complete, irreducible ideas. Where other physicists are content to radiate their thoughts constantly, in a motley, muddled mass shaped by conversation, Pauli wants to radiate only perfect quanta of thought. His genius is a mixed blessing; he can see too far ahead to proceed with abandon. He is keenly aware of all the obstacles in a path, so he treads far more carefully than a less omniscient traveler. Pauli's knowledge is almost a handicap; it keeps him from giving rein to his imagination in the way that those less familiar with "the magnificent unity of classical physics" can.

Night has fallen while I have been lost in thoughts with you and of you. I should head home now, I suppose. I still have more to say, more ideas that are swirling around in my mind, but I will write again soon—perhaps tomorrow. As it is, there is no rush to mail this letter. I couldn't if I wanted to—I wouldn't know where to send it. So, for now, I will just put it away in the tin box where all the other letters lie, waiting for you.

I woke up this morning awash in a pool of soft light. The rays of the sun were racing through the slanted windows, and I realized I had fallen asleep at my desk last night. As I attempted to shake the stiffness from my neck and shoulders, I berated myself for falling asleep at the desk, but when I came back to the pension last night, my head was wound up tight as a clock. I couldn't sleep for the whirring of the wheels and cogs in my mind, so I sat at my desk and read and thought until the steady tick-tock of ideas rocked me into oblivion. Why was I so keyed up, you ask?

I ran into Hans, Paul, and a few of the others on my way home last night, and they convinced me to join them for a movie. We went to the pub for dinner afterward, and as always, the conversation turned to physics. I don't often think about the atomic model these days, but having written to you last night, it was fresh in my mind, and we began talking about the uneasy origins of quantum mechanics. Hans said something about how, until just a few years ago, people hung onto classical laws as if for dear life. Even when it was known that nature is undeniably discrete at the atomic scale, we still tried to treat it as continuous whenever we could.

For about fifteen years, classical and quantum laws were used alongside each other, despite the fact that they were known to be contradictory. The Bohr-Sommerfeld model of the atom acknowledged that the orbits were quantized, but their energies were nonetheless computed using classical rules. The calculations worked out, but physicists were painfully aware of the glaring logical contradictions. Sir William Bragg is quoted as saying that he used the classical theory on Mondays, Wednesdays, and Fridays and quantum theory on Tuesdays, Thursdays, and Saturdays. Presumably, on Sunday, the day of rest, one can seek refuge from this infernal choice. But one can live a double life only so long, and the need for a single consistent theory was glaringly apparent.

With such drastic changes afoot, it was only natural to wonder where, in the confusing quantum world, lies our familiar classical corner of reality. Bohr addressed this issue through his Correspondence Principle, which says that, in the limit of large quantum numbers, we should be able to recover classical rules from quantum theory. Just as Newtonian mechanics pops out of Einstein's relativity in the limit of small velocities, Bohr said that when the number of quanta involved is very large, the dictates of the new mechanics should echo the statutes of old. In other words, the classical picture was not to be dispensed with entirely; it would be subsumed in the new quantum theory as a limiting case, confined to a particular corner of validity.

We were discussing the Correspondence Principle in the pub last night, when the conversation fizzled out. A group of pretty young girls walked in, and with the unspoken understanding that comes of listening to each other's thoughts all day, my friends immediately dropped their philosophical ramblings and began to vie for the girls' attention. They often tease me about my reserve in such matters, and I take their wisecracks with a smile; for how, without outright lies, or sounding insane, do I explain about you? How do I say that your absence takes up so much space in my life, it leaves no room for anyone else's presence?

I took my leave after dinner, but my mind was still feverish with unresolved thoughts when I got home, and sleep was only a minor interruption. As soon as I woke up, the same ideas crowded into my head, like loyal customers waiting for the local café to open so they can get their breakfast pastries and coffee. Within minutes, every table in my mind was full, and I was fighting to keep up with the early morning chaos and the excited idea chatter.

After a hurried breakfast, I decided to step out into the beautiful morning and let the wind clear my mind. Birdsong is more joyous now; the bulbs have begun to bloom, and the papers are full of spring poetry. Slowly the days are getting longer, extending a little at both ends like a child lazily stretching out after being curled up for a long sleep. Most everyone here is crazy about the snow, but for me the best time of year is now, when the last of the snow has melted and there is a newness in the air.

Walking our way into a solution is practically a tradition at the Institute. Whether it is through the tree-lined paths in Fælledparken, along swan-filled lakes, or on city streets, whether we prefer to be immersed in a people-bath or cleansed with nature, we all seem to have picked up the habit of thinking as we walk. This aspect of our culture, too, can be traced to Bohr. Collaborators, new initiates, and special visitors are all taken on long walks, perhaps to Klampenborg forest on the outskirts of the city, so they can talk among the beech trees, or perhaps through the island of Zealand to Hamlet's alleged castle, Kronborg.

Most of the time, I prefer urban walks, but this morning I felt like gazing on the water, and so I made my way to Nyhavn. All year round, the colorful facades on the waterfront are harbingers of spring. I walk here in the winters, when darkness falls early and I need to be reminded of the flowers to come; and when, in spring, that promise is realized, I like to visit the cheerful buildings again to acknowledge that they were right in keeping the faith.

After taking in the bustle and merriment of Nyhavn, I walked along the harbor to the quieter pier of Langelinie where I sat down. As the blue expanse washed the tiredness from my eyes, I was reminded of Bohr's remark that part of infinity seems to lie within the grasp of those who look across the sea. Absently I picked up two pebbles and threw them into the calm water, watching the ripples as they expanded, interacted, and faded away. Interference is exclusively a wave phenomenon. In fact, it is one of our litmus tests for distinguishing a wave from a particle. Two waves can interact such that they reinforce each other or cancel each other out; two particles can never do the latter. Another feature unique to waves is the ability to diffract—to fan out, so to speak, as they are forced through narrow slits. It is because light exhibits both these qualities that for centuries we categorized it as a wave. And yet, as we have learned in recent years, there are times when light simply must be interpreted as a stream of particles. Could it be that this division is not as absolute as we thought? If waves can sometimes act like particles, can particles not sometimes act as waves? While that seems like a perfectly logical step to take, it is so at odds with experience that no one thought about it until 1926, when the French aristocrat Louis de Broglie finally articulated the question.

The answer was rather illuminating. While the mathematics involved was simple enough, the ramifications were enormous and far reaching. Bypassing the entire discussion of the atomic model, de Broglie went back to the beginnings of quantum theory and the existence of light quanta. He found that a wave could indeed be associated with every material object. This has not so far been apparent to us, because objects at everyday scales have exceedingly tiny wavelengths, and consequently their wave nature is somewhat suppressed. Smaller objects—such as those with which quantum theory concerns itself—have long wavelengths and hence exhibit overtly wavelike behavior.

Perhaps his wartime work with radio waves, and his love of chamber music, influenced de Broglie to see the atom almost as a musical instrument, with the smallest, lowest-energy orbit corresponding to the basic tone, and the higher orbits to a sequence of overtones. Just as a string of a certain length can sustain only a particular set of notes, said de Broglie, the same is true of electrons in an atom. Bohr's "stationary states" are simply orbits that accommodate an integral number of wavelengths, thus allowing the electron-wave to seamlessly join back onto itself—anything else would lead to destructive interference and the wave would decay. The mysterious quantum numbers, which seemed to be arbitrarily associated with electron orbits, emerged naturally from de Broglie's model as the resonance condition for standing waves.

This newly discovered duplicity of nature, whereby matter and waves were two different sides of the same coin, was difficult to accept. Some even brushed it aside, referring to it as *la comédie française*. After all, if matter could be smeared out into waves, what were these waves made of, and how were they to be interpreted? How is it that a single object can exhibit both particle and wave behavior when centuries of experience tell us that things are either one or the other?

Faced with the contrariness of quantum mechanics, our entire classification system crumbles. The concept of location—so instinctive to our thought—no longer makes much sense in the context of an atom. We are used to thinking of orbits as paths traced by moving objects, like planets. At any single instant in time, a planet has a definite location. An electron in a nucleus is not so well located. Though we use the same word to describe it, the relationship between an electron and its atomic orbit is not that of a spatial path being traversed linearly in time. Unlike particles, waves are not restricted to a point—they are present everywhere along the oscillation. As a wave, the electron exists at every point along the orbit, like a fuzzy, nebulous distribution. It is as if the essence of an electron oozes out the tiny ball-like particle we intuitively picture.

Walt Whitman might as well have been the spokesman for quantum mechanics when he wrote

Do I contradict myself?
Very well then I contradict myself.
(I am large, I contain multitudes.)

Does nature really contain self-contradictory multitudes, or is that perhaps a consequence of the paradigms we have imposed on it? Could these puzzling aspects be complementary rather than contradictory?

Quantum mechanics' resident philosopher, Bohr, undertook to navigate this new fork in the road. Perhaps, he said, particle

ONLY THE LONGEST THREADS

and wave are complementary descriptions of a single object, but we are limited to observing only one at a time—like water, for instance, which we may perceive as solid, liquid, or even gas, depending on the conditions. The same object can take on all these guises, by turn. It cannot freeze and flow at the same time, but in the correct conditions it can do either. In similar manner, our experiments single out either the wave or the particle representation of an object. The contradiction dissolves if we accept that there is more to nature than we can directly experience. There are things so large they can have more than one truth.

Sitting at Langelinie today, I gazed out at the sea. Perched on a rock a few feet in front of me was a small statue immortalizing Hans Christian Andersen's Little Mermaid. Her heart was on the land, her body belonged to the sea, and her spirit was too vast to be fully contained by either. But the world was not equipped to accommodate both her realities, and so, to disastrous consequence, she was forced to choose.

After Bohr's proposal that particles and waves are complementary descriptions of reality, his protégés, Heisenberg and Pauli, were even more convinced that a radical reformulation of quantum mechanics was in order. I am not entirely sure that Bohr disagreed. He recognized the limitations of language and the inherent inability of our vocabulary to encompass the atom; the words created to describe classical concepts do not always fit quantum situations. In this realm, we can use language only as the poets do, to "create images and establish connections."

Imprecise word pictures did not satisfy Heisenberg and Pauli, and merely tweaking things here and there did not suffice either; they were convinced the theory had to be recast entirely. Together, Heisenberg and Pauli argued for a formalism that dispensed with pictorial representation. Almost the only thing to which they did not take exception was the fundamentally discrete nature of the quantum.

Heisenberg, expert skier, left-handed ping-pong champion, and piano player extraordinaire, stepped into the spotlight with a bang. He suggested a completely new philosophy of physics in which theories would avoid mentioning quantities that were not experimentally accessible and instead be formulated strictly in terms of observable quantities. Perhaps this idea arose from a seed planted in his mind when he heard David Hilbert lecture on mathematics, a moment when he realized for the first time that "one could have axioms for a logic that was different from classical logic and still was consistent." Just as geometry could be formulated in a perfectly self-consistent and sensible manner based on axioms other than Euclid's, and just as in relativity one could use the words *space* and *time* "differently from their usual sense and still get something reasonable and consistent," Heisenberg wondered if physics could be described by using a new scheme that differed both logically and axiomatically from classical theories and yet was consistent.

In essence, Heisenberg said, we should control the urge to connect the dots and paint a picture of what goes on behind the scenes. All we need to do in science, he said, is to predict what will happen next. We will never know for certain what goes on in the dark places hidden from view, so we should limit our reflections to those things we can directly experience, for that is all we can ever know is for real. Everything else is conjecture, and that Heisenberg had no patience for. Heisenberg was adamant that we do away completely with analogies and mechanical models, focusing only on what is distinctly measurable; in other words, that we get out of the way of the theory and resist imposing our own logic on it. That is a blow not only to the human imagination but also to the human ego.

Nevertheless, Heisenberg persevered, trying to see mathematically what was invisible to the eye. He stopped visualizing atomic orbits entirely, whether as paths traced out by little ball-like electrons or as standing waves, and instead replaced these pictures by an equation that was an appropriate abstraction of Newtonian laws. He restricted himself to considering only physical variables, like the positions and velocities of electrons.

The challenge was to see if this new formalism encapsulated enough information to reproduce atomic spectra: those patterns of frequencies and intensities that were the only concrete information we had from the atom. The existence of orbits had initially been inferred from the discrete nature of atomic spectra, but observations had gradually drifted into the background as discussion centered on deduced realities. Heisenberg wanted to refocus attention on atomic spectra and, in order to do this, he needed to find a way of describing the radiation emitted by electrons as they transitioned to different states.

Thus far, no one had ever been able to calculate the instant when a particular electron would decide to leap between orbits, but statistical laws were found to work surprisingly well. The behavior of a group of electrons could be predicted to stunning accuracy even when no claims could be made regarding an individual electron. This ambiguity troubled many, but Heisenberg decided to embrace it. He wove statistics into the structure of the theory by choosing to deal in probabilities instead of certainties. Heisenberg formulated the problem in terms of functions that measured the likelihood of a given electron transitioning from one specific state to another. Transitions with a higher probability would happen more often, and the corresponding lines in the atomic spectrum would thus be more intense.

Trusting mathematics to lead the way, Heisenberg followed his equations. He had not gone very far before he realized, much to his dismay, that the familiar rules of algebra did not seem to hold anymore. The quantities he was working with displayed a strange property that completely violated experience: the product of these quantities depended on the order in which they were multiplied. Despite his resolve to eschew preconceptions, Heisenberg found this idea hard to swallow, and he feared it signaled a fundamental problem with the theory. Luckily, the noncommutative nature of multiplication did not cause everyone to feel the same dismay. Max Born at Göttingen recognized these "new" laws as the well-known rules that had long been used by mathematicians to multiply array-like objects such as matrices. So, all Heisenberg was saying (without knowing it) was that quantities such as position and momenta are represented on the quantum scale by entire arrays instead of single numbers. Granted, this was a strange concept, but at least it did not defy mathematical logic. As a somewhat unexpected boon, Pauli gave this work his blessing, calling it the *morgenrote* (predawn) of quantum theory, and showing that it could reproduce the energy spectrum of the hydrogen atom.

At this stage of its development, quantum mechanics was beginning to sound like a complex piece of music, sung in many parts. It incorporated the voices of the intuitive as well as the rigorous; passionate mathematicians as well as visual thinkers; philosophers as well as lovers of formalism; experimentalists as well as theorists. Although the polyphony made for a richer sound, the different melodies were not always harmonious, and some notes were decidedly discordant. But, looking back, I cannot see how quantum theory could possibly have progressed this far, this quickly, had it not been tossed back and forth so many times among such different voices.

As the performance progressed, battles occurred not only over dissonant notes but over far larger issues, like chords and scales. While Heisenberg was still playing his strikingly modern, and somewhat jarring, composition, an older Austrian virtuoso wandered on stage. Erwin Schrödinger played lilting melodies with undertones of familiarity, which evoked nostalgia for the classical ways of old. These two maestros, pitched against each other, personified the epic confrontation between images and equations. It seems as if the two go hand in hand, like waves and particles; perhaps mathematicians and visionaries see different aspects of nature, and it is only by piecing their complementary images together that we can complete the puzzle.

Heisenberg's matrix formulation of quantum mechanics was something Schrödinger rebelled against, claiming to be "discouraged, if not repelled," by what seemed to him to be some technique "of transcendental algebra, defying any visualization." Schrödinger took his cue from de Broglie's wave-particle duality. Since objects at the atomic scale are small enough to have appreciable wavelengths, Schrödinger argued, quantum mechanics should be formulated as a theory of waves rather than of particles. To lend support to this view, he drew an analogy with optics. There are times when we talk about light, ignoring its wave nature completely; reflection, refraction, and the formation of images in lenses can all be understood perfectly well using ray optics alone. But light also exhibits phenomena like diffraction and interference, and the wave description must be invoked to understand these. Perhaps mechanics also works analogously, said Schrödinger. Ordinary mechanics, much like ray optics, can be used to understand classical phenomena, but it is not the whole answer; for certain other phenomena, the only possible explanation stems from an analogue to wave optics: what he called wave mechanics.

If such waves existed, we would need to know how they moved and evolved. Just as Newton's laws of motion dictate the motion of particles, a wave equation is needed to trace out the paths of waves, and this is precisely what Schrödinger found. He was able to write down a dynamical equation whose solutions had definite energies and exhibited wavelike properties. These so-called wave functions were three-dimensional and consequently could move in, and fill, the entire extent of the atom. One could say that where de Broglie played on strings, Schrödinger's musical instrument of choice was more akin to a cymbal, vibrating throughout its being.

Schrödinger's wave mechanics enjoyed its share of success, just as Heisenberg's matrix formulation had done. It appeared there were two, equally valid, vantage points from which quantum mechanics could be viewed, and this added perspective enriched the picture. Schrödinger's approach explained all the atomic phenomena encompassed by Bohr's theory as well as things like the intensities of spectral lines, which the earlier model had not been able to address. As an added bonus, it also predicted new phenomena, such as the diffraction of an electron beam, which had not even been hinted at before.

To physicists smarting from the sharp edges of Heisenberg's mechanics, the blissfully familiar mathematics of waves was like balm on their wounds. But despite appearances to the contrary, the nature of quantum mechanics was not quelled. There was a problem with the interpretation of Schrödinger's waves. These were not physical waves propagating in a material substance, but instead abstractions through which we could follow the propagation of a chance. Even Schrödinger's apparently classical methods could not coax any definite answers out of nature. We were still handed out probabilities; the only difference is that these could be read off simply from the shape of the wave function. One could set up Schrödinger's equation for an electron in an atom, but the solutions of this equation would not be definite trajectories à la Bohr; they would merely predict the likelihood of an electron being at a particular place at a particular time. The orbits of old, inked onto the page in clear strokes, had been smudged and smeared.

It began to seem that the landscape of reality was painted in watercolor: paints bleeding, diffusing, melting into each other, thinning out to the extent that you can't say where they end. Our natural inclination is to interpret this as the spreading out of concentrated particles into nebulous blobs of diffused matter. If only it were that simple! In Eddington's words, "The spreading is not a spreading of density; it is an indeterminacy of position, or a wider distribution of the probability that the particle lies within particular limits of position. Thus if we come across Schrödinger waves uniformly filling a vessel, the interpretation is not that the vessel is filled with matter of uniform density, but that it contains one particle which is equally likely to be anywhere." Our watercolor landscape is at best an allegory of reality; it is definitely not a photograph. There is very little here that we can take literally. No wonder Eddington warns that it would "probably be wiser to nail up over the door of the new quantum theory a notice, 'Structural alterations in progress—No admittance except on business,' and particularly to warn the door-keeper to keep out prying philosophers."

The loss of determinism was disconcerting, to say the least. Everyone struggled to make sense of this unexpected universe we had unknowingly inhabited for so long. One of the best expressions of intellectual frustration came from Cambridge University, in the form of an anonymous poem some students tacked on the walls of Cavendish lab. It was a "petition" made by electrons desperate to be set free from the "dread uncertainty" of the "hated quantum view." They lamented the loss of the "smooth-flowing" time when all they had to do was follow the classical equations, and they bemoaned the sudden identity crisis they found themselves in, no longer knowing if they were particles or waves or a "jelly sort of phi," or even if they were real at all, "or where we are or why."

The nostalgia for a classical deterministic past was felt by many, but a stronger source of frustration was the realization that we just could not describe quantum reality using our existing vocabulary, even if we combined words in radically new ways. As Schrödinger said, the leap quantum mechanics required of us was not the transition from ordinary lions to "winged lions"— which, though not in our experience, are at least imaginable— but instead something akin to the leap from circles and triangles to apparently self-contradictory entities like "triangular circles." He went so far as to say that such a model was not even thinkable. And yet, what can one do except to think?

As yet another day passes, I think of all this and I think of you. More and more, you are becoming inextricable from my thoughts of physics. These are the ideas that flow through my veins right now, and the only reason I am putting them down is to share them with you one day. Once you read them, you will come to know me just as well as if you had been at my side all along.

Around the time Schrödinger was working on his wave equation, Heisenberg came up with the Uncertainty Principle—his masterpiece. Heisenberg discovered that, for a pair of non-commuting variables (like position and momentum, or energy and time), a simultaneous determination of their exact values is impossible. There is an inherent uncertainty in both the measurements, and the product of these uncertainties is a constant—Planck's constant, it was called. This simple equation lays out an explicit mathematical condition, to be satisfied in any way we choose. We can allow a small uncertainty in the measurement of both variables, or we can pin down one of the quantities exactly while leaving the other undetermined; either way, the more accurately we know the value of one quantity, the less certain we are about the other. Understandably enough, people had trouble swallowing this new principle. Even Pauli wrote to Heisenberg in exasperation, asking him to help make sense of the fact that "one may view the world with a p-eye and one may view it with a q-eye, but if one opens both eyes at the same time one goes crazy."

This ridiculously unintuitive statement actually makes sense if we try to picture it in terms of particles and waves. The only way to constrain a wave to a precise location is by adding to it a large number of other waves, of varying wavelengths, in such a way that they reinforce each other close to a single point and cancel each other out everywhere else. This results in a sharply focused "storm" that resembles a particle. Since it is localized, the position of this "particle" can be accurately determined, but it no longer has a specific wavelength.

Alternatively, we could choose to determine the wavelength exactly and restrict our beam to a single wavelength, but then the position is no longer precise, since a wave cannot be said to exist at any one point; it must of necessity extend over a region of space. So, we can know either the position or the wavelength (momentum) of an object with increasing accuracy, but each choice necessarily introduces an ambiguity in the measurement of the other quantity.

Mathematically, this tug of war is expressed by saying that position and momentum do not commute. If they did, both could simultaneously be precisely determined. As it turns out, they *almost* commute, missing each other only by a minuscule amount, and the gap between them is precisely the size of a quantum. So, in a sense, the quantum measures the uncertainty inherent in our measurements.

While this uncertainty is rather small, its philosophical implications are huge: either one can never know a state exactly, or worse—a state with both a definite position and a definite momentum simply does not exist. Although scientists had never succeeded in specifying all the parameters of a physical system with perfect precision, the assumption remained that it was possible to do so, but the Uncertainty Principle deals a deathblow to that long-cherished belief. As Paul Dirac put it, when we claimed to be able to predict the future based on the present, our conclusion was incorrect, not because the logic was faulty but because the premise was false. We are barred, as a matter of principle, from knowing the present completely.

Quantum mechanics not only rids us of determinism, it also places the responsibility squarely on our shoulders. Our decisions and measurements influence a quantum system and change its reality. Apples and elephants and trees are unaffected when we see them, because the impact of a single light ray on an apple is insignificant, but this is not true on subatomic scales. If we want to see an electron, we shine light on it, thereby disrupting whatever it was the electron was doing. By choosing to spy on the electron, we change its behavior. Not only is the universe a far more bewildering place than we ever imagined, but it is no longer completely external to us; somehow we have ended up with the controls in our hands. This confusion is vaguely familiar. It feels like growing up.

Debates about quantum mechanics were impassioned and desperate, and quite unlike the civilized conversations in which earlier generations of physicists indulged. On one occasion, Bohr questioned Schrödinger so ceaselessly that the latter succumbed to fever from exhaustion, and even then Bohr continued to argue by his bedside. Nothing was held back as friends and colleagues battled it out. Tongues flashed and tempers flared. Hans claims to have seen a letter to Pauli from Heisenberg, saying that if he was ruining physics, it was not out of malicious intent, and that if he was being reproached for being such a big donkey that he never produced anything new, then Pauli was a jackass also, because he had not accomplished anything better.

Amid their quests and duels, these zealous knights had dark periods of doubt and despondency. Frustrated at not being able to rid the theory of its "damned quantum jumping," Schrödinger said he regretted ever having anything to do with it, and Pauli expressed a wish to become a movie comedian instead of a physicist so he could be spared the whole drama. On a more serious note, the mental demands the theory made were unprecedented, and everyone felt the pressure. Max Born had a nervous breakdown from the strain of his efforts to "keep up with the young ones," and many of quantum mechanics' main actors bemoaned their complete and utter incompetence, their lack of even a modest grasp of the theory.

To add to this morass of confusion, probability, and uncertainty, there still existed two entirely different pictures of quantum mechanics—Heisenberg's matrices and Schrödinger's waves—and no bridge connected the two. None, that is, until the wave-particle spirit channeled itself through a young English physicist, to show once again that quantum mechanics was not an "either/or" game but an "and" game. It was large enough to contain two apparently disparate realities; it was more than both.

As the details of matrix mechanics and the interpretation of probability waves were being ironed out on the Continent, Paul

Dirac was quietly at work in Cambridge. In many ways, Dirac was an anomaly among his loud, larger-than-life contemporaries. Dirac was fascinated by beauty, but he had little time for literature or the theater and thought philosophy a waste of time; he sought, and found, a majestic beauty in mathematics. Dirac thrived on its enduring nature and its regal lack of ambiguity; there is no room for dissent, or even opinion, regarding a mathematical truth: it is absolute and eternal. Perhaps it was because of this preference that Dirac worked alone. Words crave company, they thrive on conversation and interpretation, whereas mathematics has its own internal checks and is free of the need for external validation. Like mathematics, Dirac was content unto himself. When he chimed in to the quantum mechanics chorus, Dirac attracted immediate attention. His notes were so pure and so sure that they cut through the hum, and all around him stopped to listen. The song he sang was a hymn to mathematics.

It was clear to Dirac that, given the current state of quantum mechanics, a fundamental change of formulation was required. Since stable theories are erected on solid, powerful, mathematical foundations, Dirac set out to revolutionize the underlying mathematics of quantum mechanics. In his opinion, the crucial feature of matrix mechanics was that it involved noncommuting variables, and he wondered if such objects could be encompassed somehow in the familiar structure of dynamics.

Using a century-old formulation discovered by William Hamilton, which could rather straightforwardly be generalized to noncommuting variables, Dirac was able to express matrix mechanics in a form reminiscent of classical dynamics.* Once this esoteric theory had been expressed in a familiar manner, it was relatively simple to see that the equations of Heisenberg's theory, and the vastly different wave equations put forth by

*With the understanding, of course, that the physical variables no longer commute.

Schrödinger, could in fact be transformed into each other and hence were equally valid and in fact interchangeable.

This might be the most powerful lesson quantum mechanics teaches us: that the whole is more than the sum of its parts. Things may appear distinct on the surface, but when we probe deeper, we find they are merely different aspects of an underlying reality that is more complex and wonderful than we can fathom. Maybe nature's true visage is so stunning that we can bear to see only filtered images. Each filter highlights unique features, all of which must be combined to create an impression of the whole. If this is indeed so, perhaps our most valuable contribution to science lies in the individuality of the filter we impose on the world.

For perhaps the first time, physics has entered a realm where interpretations are unclear. Our subjectivity is as valuable as our objectivity; what we see and how we perceive it are equally important. In some ways, these physicists puzzling over the quantum might as well be so many artists painting a rose. The work of each is different, and true in its own way, but the artist cannot record the rose without also revealing his own impression of it. And should the rose happen to be hidden from sight, the more perspectives and insights we have, the better.

In Victorian times, physicists said they truly understood something only if they could construct a mechanical model of it. But at the frontiers of knowledge where we stand today, mechanized models are a far cry—even language breaks down. All we have left are symbols. Eddington writes that the world of physics is becoming increasingly symbolic, and so now it must be built by mathematicians, rather than engineers, for these are the "professional wielder[s] of symbols." But as even he admits, "It is difficult to school ourselves to treat the physical world as purely symbolic. We are always relapsing and mixing with the symbols incongruous conceptions taken from the world of consciousness. Untaught by long experience we stretch a hand to grasp the shadow, instead of accepting its shadowy nature."

Of all that quantum mechanics requires from us, this is perhaps the most difficult task to perform. We are so used to visualizing, to imposing a familiar narrative on things, that it feels strange to manipulate symbols that lack concrete meaning, using rules that are self-consistent but wholly foreign.

Then again, far be it from me to disparage the use of symbols, for is that not what you are? A symbol for my hopes and dreams, the missing abstraction that makes sense of my life. If I postulate that somewhere in this vast world you exist, everything falls into place. Just by assuming that, somewhere, there is a mind which vibrates in sympathy to my own, my thoughts are validated. You are not a figment of my imagination as much as an invention of my need. Whether or not you live and breathe, you make my ideas come alive, and so, until I am faced with proof to the contrary, I choose to believe that you exist. I even dare to dream that, one day, I will meet a girl who is a tangible manifestation of the abstract idea that is you.

I came back to Auditorium A to finish this letter. Sitting here now, I lift my gaze again toward the photographs on the wall. What a Pandora's box these men have opened! They have dispensed with determinism, reached beyond experience, mixed up waves and particles, blurred the boundaries of what was defined, and chopped up nature into miniscule pieces. Even after all these years, the stinging graininess comes as a strange surprise, as if you found sand under your feet when you expected marble.

Looking up, I realize how dark the room is getting. Twilight is melting slowly into night. I can no longer make out the equations on the blackboard, but even now I feel their pull. How potent they are, these scribbled symbols, these dim one-dimensional projections of a multifaceted reality! How intensely they draw you in, how much discipline and loyalty they demand. Time and time again, equations have proved that if you are loyal, and follow the circuitous paths they trace, their ways slowly become clearer.

Ridiculous, that one would need to pledge such unwavering allegiance to chalk dust—and yet the rewards they bestow on the faithful are so tempting, so full of wonder, that one cannot help but be seduced. So I sit with them in the dark another few minutes, letting their wave functions intermingle with mine, hoping for constructive interference.

Email: Sara to Leo

From: Sara Byrne <breaking.symmetries@gmail.com>
Date: Wed, Jan 23, 2013 at 4:56 PM
Subject: Through Unimagined Ways
To: Leonardo.Santorini@gmail.com

Dear Leo,

A biting wind prowled the streets today, while I snuggled up indoors with the next installment of your manuscript. I love the way this is going. After having studied these theories, rigorously, for years, I really appreciate having the familiar made strange. It's such fun to look at the world through alien eyes, to sift ideas through the sieve of a foreign mind. Just by seeing the same things expressed in different words, I am picking up on nuances I had never before noticed.

A thought experiment was, to me, a clever workaround for the difficulties of practical implementation. By turning the mind into a laboratory, you extend the realm of what is accessible. No longer bound by physical constraints, you can "experiment" on anything that is conceivable. At least, that's how I always saw it before. But when I read Jacob's descriptions, I realized that Einstein had developed a feel for things by thinking himself into the picture. His *gedankenexperiments* weren't carried out from a safe distance; he was in the hot seat.

Einstein didn't frame his questions as: "what happens to light if . . ." but instead as "what would *I* see if I were to ride a light beam?" or "what would *I* feel if I were in free fall, in a box out in space, far away from any gravitational field?" Einstein cast himself as the protagonist of the story. That's exactly what we do when we read fiction! I had never noticed the parallel before.

On another note, I totally agree with you about the toll quantum mechanics took on many brilliant minds. One can't help but sympathize. Classical physics is so much more in tune with the way our brains are wired. Even the iconoclastic theory of relativity is comprehensible. It might stretch your mind to the elastic limit, but you *can* wrap your head around it. To accommodate quantum mechanics, you pretty much have to change the topology of your brain.

In spite of all the mayhem it caused, quantum mechanics did relax our attitudes. In years gone by, people treated work with a ritual respect. Physics was a lofty calling; it even inspired religious feelings or aesthetic pleasure. With the advent of quantum mechanics, that formality began to fade. The culture became more fun. There was room for whimsy, even irreverence. Maybe this shift was a global sign of the times and had nothing to do with physics—but I can't help thinking quantum theory was responsible. How formal can you be with something that crazy?

You wanted my opinion on the explanations and analogies you use. For the most part, I think they work pretty well, but they can't possibly be perfect—that's just the nature of the problem. The only way to do these ideas justice is to express them mathematically, but that's a pretty intense undertaking and doesn't fit the spirit of this book. I suppose what I'm saying is, go ahead and use your analogies, but know that they will all break down somewhere. Sooner or later, someone will take issue with each one. Still, using them is better than not. They're all we've got to go on. Personally, I think the more, the better. And in that spirit, here's one to get you going:

Metaphors are like sheets of paper. Any flat surface can be covered by a single sheet of paper, without wrinkles or folds. To completely conceal a curved surface, you have to glue down several small sheets, often overlapping. Simple ideas, like flat surfaces, can be covered by one metaphor. The advanced ideas you're dealing with are highly curved—no single metaphor can span the entire surface. You have to work in patches and use a different simile for each aspect of a multifaceted idea.

You asked whether I've thought about writing the string theory chapter. The truth is, I'm a little apprehensive. I'm not sure I could do the subject justice. It's like asking an art student to make a reproduction of a painting by a grand master. What if I mess it up, and because of me, people turn away from a genuine masterpiece? It's a scary proposition, but I promise I'll think about it over a cup of hot chocolate.

That probably sounds strange, but I'm sitting by the window, and every time I look up, the brown and pink awning of Burdick's beckons. I've held out for hours, and I'm finally done fighting temptation. Not that I need an added incentive, but the shop seems quite empty right now. Actually, everything is, Gutman Library included. Most of the large, comfortable chairs around me are unoccupied, except by the light that floods in through large glass panes. It won't last. In just a few days, winter break will be over. People will be scrambling for chairs in the library, fighting over every inch of parking space outside, and waiting in line for Burdick's liquid ambrosia.

But today I have Cambridge almost to myself. So I'm going to grab that hot chocolate and go on a long, leisurely walk along the unbelievably quiet streets of Harvard Square, to drink in this peace and quiet before it melts like the snow.

Sara

P.S. As far as I'm concerned, that whole Anna thing totally works. I don't know how many will agree with me, but I loved the twist!

The Third Installment

From: Leonardo Santorini <leonardo.santorini@gmail.com>
Date: Tue, Mar 26, 2013 at 12:03 AM
Subject: The Third Installment
To: breaking.symmetries@gmail.com

Dear Sara,

Here, at long last, is the final installment of our book. I say *our book* because I am not sure I would have plucked this idea out of the void on my own. At times I am convinced there's an idea field pervading space, and that some interactions have just the right combination of energy and serendipity to shake an idea loose. If that is indeed so, the spark in our conversation last July did the trick. It got me started. But I owe you more just than the beginning.

Writing this book has been equal parts delight and torment. Had you not been there to talk it over with, I would have abandoned the project a long time ago.

I still think a string theory chapter would finish things off nicely, but I leave that up to you. It would be fabulous if you wrote it, because these are ideas you have actually lived with. You will relate to them in ways I won't. I don't mean to push, but if you decide you don't want to write the chapter, would you maybe consider sketching it out and getting me started? Think about it and let me know. No pressure. You've already helped more than I thought possible.

Bohr and Abdus Salam knew what they were saying: ideas do crystallize in conversation, and research cannot be conducted in a vacuum. One needs a sparring partner. You have been far more than that for me. Each time I tossed you an idea, you sent it back to me larger, more animated, in deeper color; this book is really yours as much as mine. I have said my piece, and so it is over to you. Take the manuscript and do with it what you will; pick it apart or fill it in, whatever feels right.

Ciao,

Leo

P.S. Happy birthday. Keep an eye on your mailbox.

CHAPTER 5

The Greatest Forms of the Beautiful

[Particle Physics and Electroweak Unification]

The mathematical sciences particularly exhibit order,
symmetry and limitation; and these are the greatest
forms of the beautiful. —ARISTOTLE

JUNE 1981
TRIESTE, ITALY

My darlings, Fatima and Hassan,

This letter is only for the two of you and, as promised, it is a
"proper" letter, addressed not to the children who sat on my
knee but to the young adults you are now becoming. I think
back to those days (which doubtless seem long ago to you, but
are only yesterday to me) when you would leaf through my papers, fascinated by the fact that I could "read" the funny, squig-
gly alphabet of equations. I remember how eager you were to
learn to draw those symbols, how you traced their foreign
shapes on page after page . . . almost like some rite of initiation.
"What does this mean, Abba?" you would ask, and I would have
to fend you off with incomplete answers. I have waited a long
time to share with you the thoughts that occupy my days, so this
letter marks an occasion in my life as much as in yours.

It is the lunch hour, but I have skipped the long queues in the
cafeteria and crossed through the wall from the ICTP* into Miramare Park. A deep calm pervades the wooded groves, and in the

---

*International Centre for Theoretical Physics.

clearing a picture-perfect fairy-tale castle hovers on a cliff over the shimmering Adriatic Sea. On my walks here, I have stumbled on picturesque little fountains, burst on spectacular views, and in quiet places where the curtain of leaves parts, I have even glimpsed the elusive deer who call this park home. I imagine the two of you running around, uncovering these half-hidden pleasures, and the thought makes me smile. My first few days in Trieste, each time I saw a graceful statue, a beautiful building, or a lovely vista, I would turn around instinctively, looking for you. This city is replete with sights I want to share with you and your mother. I miss you all so much, it makes my heart ache. If God wills, one day the four of us will visit Trieste together.

There is one particular bench that I have come to think of as my own. From here, I can see the immensity of soothing water that stretches across to the horizon and also the hills that cup around Trieste like the palm of a gentle hand. Sitting here, far from the clamor of mindless chores and institutional responsibilities, I find my perspective on physics broadening; perhaps perspectives are elastic and stretch to fill the space provided. In this wide expanse I spread my thoughts: I lay out exciting new ideas, poured into my head during the seminars, and pull out the crumpled thoughts I had pushed to the back of my mind. Then I let all these ideas drift like clouds across my mind's sky while I sit back and watch, marveling at the pictures they form. Some of these shapes I will attempt to point out to you now. But, as I told you when you were little children, fluffy cotton-wool clouds vanish into wisps and vapor when you get close enough, and so, even while you strain to see the pictures I show you, keep in mind that with time and distance they may well change.

To begin: I told you years ago that I am a particle physicist. The particles I study are the building blocks of nature, much like the pieces in your Lego set. With these fundamental particles, as with Lego bricks, we must know their possible combinations and various attributes before we can use them to construct anything.

In the early part of this century, physicists thought our list of essential pieces and attributes was complete. Every phenomenon that had yet been observed could be ascribed to one of two interactions—gravity or electromagnetism. Three subatomic particles—negatively charged electrons, neutral neutrons, and positively charged protons—were thought to be the indivisible constituents of matter. The atom was built by placing electrons in orbit around a nucleus, consisting of neutrons and protons packed together.

This is how the atom is pictured in your schoolbooks, but if you approach it with a questioning mind—as you should get in the habit of doing—you will instantly realize that this simple model is in flagrant violation of the laws of electromagnetism. The recognition, and resolution, of these contradictions led to the inclusion of a few more particles and two additional interactions on our list. Since these new forces made their presence felt only when we started poking around at atomic scales, they were called the nuclear forces.

One obvious problem with the simple atomic model had already been resolved by quantum mechanics. Despite their attraction to the nucleus, electrons were kept at bay because they were constrained to move in orbits, and there was a smallest orbit beyond which they could never go. This stopped the atom from imploding, but what kept it from exploding? The very formation of the nucleus, in fact, went against the dictates of electromagnetism. By law, a group of protons should experience intense electrostatic repulsion, and yet somehow they were held tightly together. There was obviously a very strong binding agent at work, overpowering the protons' urge to fly apart; it was named the strong (nuclear) force.

Even with this addition, the acrobatics of the subatomic particles could not be fully explained. One particularly troubling move, beta decay, involved the transformation of a neutron into a proton and electron. The most obvious explanation was put forth: perhaps the neutron was a composite object made up of a proton and electron; beta decay would then simply be the dissolution of the bond that held these two constituents together. But, after careful consideration, this possibility was ruled out, and we were left confronting the perplexing display of a neutron spontaneously morphing into a proton and an electron. The three forces we knew of pushed and pulled and otherwise affected the motion of an object, but they did not change its identity. It became obvious that an entirely new kind of interaction was responsible for beta decay—the so-called weak force.

The number of fundamental forces had now risen to four, but the problems were far from over. In complete disregard for norms and tradition, beta decay appeared to flout the law of energy conservation. Experiments revealed that the combined energies of the electron and proton after decay were less than that of the original neutron. This was nothing short of a catastrophe. In an increasingly uncertain world, one of the few tenets physicists held was the law of energy conservation. Regardless of the nature of the interactions involved, the total energy of a system before and after any physical process was always the same. Energy could not be created or destroyed; it could only change form.

It seemed as if the axioms that had provided warmth and shelter were falling to pieces, and we were cast into cold, unbounded ignorance. At times like this, when gaping holes appear in a system of thought, there is only one thing to do. We must take the pieces apart and reassemble them to create a structure that encompasses the successes of old but leaves room for explanations yet to come.

In a way, it is similar to assembling a jigsaw puzzle, except that you have no image of what the completed picture should look like. Our dining table has been strewn with puzzle pieces enough times for you to know that the first thing to do with a new puzzle is to assemble the frame. Corner pieces are the most valuable finds, because they provide clues about two different sides and the way in which they meet. Once the frame is in place, it becomes easier to work in from the edges, or even to link pieces in random disconnected vignettes, because you know that somehow they will fit inside a circumscribed boundary; it may seem paradoxical, but the containment is liberating.

Laws are to a theory as the frame is to a puzzle. Once we determine the laws by which a theory abides, we have established the borders within which all our explanations must be contained. We move freely within this frame while remaining aware of the lines that must not be crossed. Energy conservation had always formed a crucial part of every frame, but faced with the conundrum of beta decay, some desperate physicists considered crossing this boundary. Wolfgang Pauli was unwilling to take this sacrilegious step, so he chose the only other option. Based on blind faith in the law, he claimed that energy *was* in fact conserved; the missing energy had merely adopted a form we could not detect. Pauli claimed that the neutron decayed into an electron, a proton, and an invisible something extra. He claimed that if the energy carried off by this phantom was accounted for, the total energy before and after the decay would be the same—just as the law of energy conservation decreed.

There were many who refused to believe in the existence of a smooth criminal who left no fingerprints behind; they tried, instead, to come to terms with the fact that energy had actually been destroyed. But Pauli held his ground. In true Sherlock Holmes fashion, he started listing the attributes the thief must have in order to execute this perfect crime. He found that the deed could be done by a particle that had no mass,* no charge, and carried a spin of ½. He called this the neutrino. Since most particle detectors work on the principle that charged particles leave visible tracks, it was easy to argue that an electrically neutral, massless particle would escape detection.

---

*Subsequent research has revealed that the neutrino may perhaps possess a small, but nevertheless finite, mass.

Two decades after Pauli placed this 'Wanted' ad, the neutrino was finally captured, and experiment vindicated the bold trust Pauli had placed in theoretical reasoning. This changed forever the way physics was done. The neutrino was the first particle whose existence was inferred in this manner, but it was far from being the last.

Even after it had been found, the neutrino continued to shock us with its maverick behavior; it turned out to be chiral. Before you can feel the horror that chilled the physics community at this discovery, you will need to know what chirality is.

Think of the electron as a little ball that rotates about an axis through its center. This axis can be oriented either up or down—corresponding to "spin up" and "spin down"—so you can picture it as a little arrow that points in the appropriate direction. (Keep in mind that these are just visual metaphors, the electron is not in fact a little ball, the quantum mechanical property called spin is not a literal manifestation of the English word, and these arrows are purely imaginary!)

In order to specify the motion of an electron, we have to state the orientation of the arrow (spin) as well as the direction in which the electron rotates. This combination of the two attributes is referred to as helicity or handedness.

To understand how helicities are assigned, try this: Point the thumb of your right hand up and curl your fingers around. You will see they turn in a counterclockwise direction. By analogy, a right-handed electron is one that rotates counterclockwise around the vertical axis when the arrow points up. By flipping your thumb, you can see that a right-handed electron will rotate clockwise when the arrow points downward. In similar vein, a left-handed electron rotates clockwise when the arrow points up and counterclockwise when down.

Notice that the two are mirror images of each other. If we place a right-handed electron (clockwise rotation, spin down) in front of a mirror, its reflection will rotate counterclockwise while spin remains unchanged: the mirror electron is thus left-handed. This interchange of left and right is in accord with expectations and familiar from everyday experience: if you raise your right hand in front of a mirror, your reflection raises its left. But other than this left-right exchange, we expect the mirror world to be identical to our own. In other words, we expect the laws of physics to be unchanged upon reflection; we expect nonchirality.* The other alternative is a chiral theory, in which the world and its mirror image have the freedom to act independently, and no one took that option seriously. Almost implicitly, we assumed that every sensible theory must be nonchiral. Imagine the horror physicists felt when they found that the neutrino was shamelessly chiral—it simply had no reflection! All neutrinos are left-handed. They lack a right-handed counterpart† and, hence, a mirror image.

I have tried to proceed systematically, but I have introduced so many new ideas that it would be quite natural if, in your zeal to understand each line, you have by now lost sight of the larger picture. Before I go any further, let me reiterate: Our study of atomic structure revealed discrepancies in existing theories; gravity and electromagnetism had to supplemented by the two nuclear forces—the strong force, which kept the atom stable; and the weak force, which caused particles to transmute and was discovered to be the mechanism behind the radioactive decay of nuclei. Our investigation of beta decay resulted in the inclusion of a new particle in the subatomic roster: electrons, protons, and neutrons were joined by the notoriously hard-to-pin-down neutrino, which also turned out to be chiral. This was such a re-

---

*Chiral comes from the Greek word for "hand." A nonchiral theory is "neutral"; it does not prefer any one chirality to the other. A chiral theory is biased in favor of a particular choice.

†In recent years, there has been some evidence that neutrinos may not in fact be completely massless, and consequently right-handed neutrinos may exist. Even if this turns out to be true, the left- and right-handed neutrinos will still not be perfectly symmetric, so the theory will continue to be chiral.

markable departure from expectation that it took a while, and much evidence, for the physics community to digest the information and accept it as fact.

But there was still more to come. The discovery of the neutrino was very welcome because it exonerated our theoretical structure, but it was considered to be a special case. We did not anticipate, nor did we want, the hailstorm of subatomic particles that soon pelted down on us.

You might be wondering where all these "new" particles were coming from, and why we hadn't seen them before. The answer is that most of these particles are unstable. They exist only for brief flashes of time, before nature redistributes their attributes into more efficient packages—and these stable parcels are what we see in the world around us. With the advent of particle accelerators and colliders, we were able to spy on the subatomic world for the first time. From Einstein, we know that mass is just a store of energy. If that seems too abstract a notion, think of particles as vessels. Each particle has its own characteristic vessel, of a particular size and shape; the amount of energy the vessel can hold determines the mass of the particle.

In these large, complicated machines, particles are accelerated to very high speeds, so that they acquire even more energy, and then they are made to collide. The vessels break open, and molten energy rushes out, like a genie escaping its bottle. This unleashed energy does not stay free for long; some of it adopts a less constricted form, like heat or light or motion, but most flows into a waiting set of empty containers—thereby taking on a new particle incarnation. Initially, energy tends to flow into fewer, larger vessels; but these are unwieldy, and it is instantly poured out again, distributed among the small, stable vessels of familiar particles. The intermediate step happens so quickly that we were unaware of it until advances in technology made it possible to see those transitory particles whose presence we had never suspected.

For a while it was exciting to find new particles, but the novelty quickly faded. Over the next couple of decades, the tale of discovery was told so many times that it lost its charm. Jokes were made* about awarding the Nobel Prize to the physicist who actually managed to avoid the discovery of a new particle. Indifferent to their reception, particles continued to rain down from the skies, and physicists scrambled to accommodate them somehow.

But even scientists have their limits, and when the muon was discovered in 1936, it was almost an affront. There was no reason for the muon to exist. For all intents and purposes, it was identical to the electron, just two hundred times heavier. Isidor Rabi (who later won the Nobel Prize) expressed his exasperation: "Who ordered that?" he asked.

One can easily sympathize with the plight of the physicists at the time. If particles decided to replicate themselves in ever heavier versions, there would be no end to the madness. There seemed to be no logic anymore behind the appearance of the new particles. People began to wonder which of these were truly fundamental. Were all of them entirely new, or were some of them combinations of the old and familiar?

While strolling through town a few days ago, I noticed ropes strung at various places along the roads. I remembered having seen similar ropes lining the paths in Miramare Park, and the repetition piqued my curiosity. Upon enquiry, I found that they were there as protection against the Bora, a wild, wolfish wind that howls across Trieste en route to the gulf. As it squeezes through the gaps in the surrounding mountains, the bitterly cold Bora becomes icicle sharp. On its prowl across the city, this wind rips the tiles off roofs, sends parked motorcycles flying in the air, and blows people off their feet—hence the ropes. Some-

---

*Oppenheimer is credited as the author of this joke, but I have not found the original statement anywhere, only indirect reports.

thing similar was happening in particle physics a few decades ago. A razor-sharp wind tore through the edifices of our theories, causing entire floors to cave in. New particles were blown in, and they flew every which way with reckless abandon, knocking people off kilter.

The Bora is rather mercurial; it blows in fits and starts. Even Triestinos accustomed to this erratic behavior are sometimes tricked into thinking the wind has run its course, and they move away from the ropes, only to find the Bora lying in wait. It was so in particle physics as well. Every time there was a brief lull from the storm and people moved in to survey the damage, the temperamental wind threw another tantrum. When it finally blew over, all that remained was a broken, unrecognizable mess. As physicists rummaged around in this windswept chaos, organizing and classifying particles, they noticed certain patterns emerging. No one understood where these correlations came from, but that was immaterial. Patterns in themselves are powerful and persuasive. Wherever there is repetition, there is the hope that reality can be peeled back one layer further, that an underlying structure exists on which the visible form is built.

In the end, it was the American physicist Murray Gell-Mann whose tabulation made this structure transparent. Gell-Mann found that a set of six elementary particles, which he whimsically named "quarks," could be combined in various ways to form the proton, the neutron, and most of the other, newer subatomic particles. This discovery also exposed the true nature of the nuclear forces: the weak force changed one type of quark into another, whereas the strong force was the glue binding quarks together. The strong force shackled triplets of quarks to form neutrons and protons, and it caused the quarks in neighboring neutrons and protons to attract each other with such power that the stability of the nucleus was guaranteed.

Not all particles could be built from quarks: the electron was found to be an elementary particle in itself, as were its heavier cousin the muon, a yet heavier version called the tau, and the neutrinos associated with all three. Together, the electron, muon, tau, and their respective neutrinos were referred to as leptons. Being free of quarks, the leptons were indifferent to the strong force, though they were still subject to the weak.

At the end of all this reorganization, the tally for "matter particles" had risen from three (electron, proton, neutron) to twelve (six leptons and six quarks), and the fundamental forces had gone from two (gravity and electromagnetism) to four (the strong and weak nuclear forces included). Not a huge improvement as far as numbers go, but not bad either—particularly in the wake of the recent particle explosion. With these amendments to our list of ingredients for the universe, we could once again find recipes for everything we had ever encountered, whether directly through our senses or indirectly through instruments like telescopes and particle detectors.

A lucky glance at my watch reminds me that the lunch hour is over and the conference is about to resume. I must get back to ICTP now. I will finish this conversation with you tonight.

Long after I put my pen down, I was composing this letter in my mind. Throughout the day, I kept searching for the right words, an expressive turn of phrase, an image, a metaphor—any vehicle that would convey my ideas to you. When I finally stepped out of the lecture hall this evening, the mental toil of the day had left me exhausted. Like a homing pigeon, I headed to the bus station at Piazza Oberdan.

Already, in the short time I have been here, Bus 42 has become my favorite place to relax. For an unhurried hour and a half, it traces a path out from the heart of Trieste through ever quieter streets to Opicina, a picturesque little town next to the border. As we curve along the gently winding roads, breathtaking vistas roll out steadily. We ride past the huge stone monuments of Monrupino, past the Grotta Gigante cave whose depths glimmer with crystal, past the lighthouse Faro della Vittoria where Winged Victory stands, keeping watch over Trieste.

Most of the passengers are locals who chat away happily on their ride home, the lilting cadences of their speech forming the perfect accompaniment to the lush scenery. Picturesque houses cling to tree-lined hills, bathed in rainbow hues by the setting sun as the boundaries between sea and sky melt in an other-worldly glow. In this setting, the rapturous notes and decadent melodies of opera seem inevitable; the lavish beauty of Italy calls these sounds into being.

Passengers alight every so often, and still I stay on, until we reach the top of the hill where the entire view lies before our eyes, like a luxurious silken tapestry unfurled. This being the end of the route, the driver parks the bus. He looks at me questioningly, but when I nod no, he smiles and heads out for a quick break. About ten minutes later, we begin our descent back to town.

It is dusk now, and the hills are dappled in pinpricks of light; houses, shops, and villages are represented by constellations of tiny dots. Instinctively I start looking for a structure, a way to group them together and make sense of them. How dense would the lights have to be to constitute a village? What might cause those outlying patterns? Are there any specific shapes I can identify that would help me recognize a specific grouping again? While my mind idly plays these games, the particulars fall away and the larger theme I have been unconsciously grappling with all day shines into view.

The mathematical meaning of symmetry is not far removed from the meaning ascribed to this word in common parlance. If I ask you whether or not a given object is symmetric, you will most likely have an immediate answer. Even without conscious consideration, you respond instinctively to geometric similarities between two aspects of the object—if such similarities exist. You might need to focus a little harder to identify and describe the symmetry, but a first glance suffices to recognize that a repetition of some kind is at play.

More often than not, symmetry is pleasing to the eye, and as a result it is heavily employed in the ornamentation and design of traditional objects. The aesthetic value ascribed to symmetry has been a constant through centuries of changes in taste and fashion. Despite all that divides the two of you from the great Mughal emperors, you have this in common with them: the intricate motifs that tessellate the arches and ceilings of the Badshahi Mosque take your breath away; you are fascinated by the geometric fretwork of the marble lattices in the Lahore Fort; and you are delighted when you come across a familiar pattern, repeated on a different scale, in an alien context or on an unrelated structure.

Being an ardent admirer of Mughal architecture, I have often thought about why I find these designs so appealing, and I think it is this: patterns are reassuring. They promise the return of the familiar, and so they hold us safe. Yet patterns have no natural end. Iterations can go on forever, and in that way the known carries us into the Unknown, offering a hint of the divine. Perhaps this is why a devout believer like Abdus Salam* was drawn to symmetry like a moth to a flame.

Inherent in the word "symmetry" is the implication that some operation can be performed on a system without affecting the outcome. A mirror-symmetric shape, for instance, can be flipped over and both the object and its mirror reflection will appear unchanged. An equilateral triangle can be rotated about the center, and as long as its three vertices fall on the same three points, even the most careful observer (as long as he has not seen you perform the rotation) will never know that anything happened. Similarly, a square can be rotated by a right angle— or two, or three, or four—without anyone ever knowing the difference.† The ultimate symmetry, of course, is that of a circle: it

---

* Abdus Salam was a Nobel laureate and the founding director of the ICTP.

† In general, a regular n-sided polygon can be rotated through any of the angles that result when a circle is cut into n equal slices, and none of these manipulations will leave behind a trace.

The Greatest Forms of the Beautiful

can secretly be rotated through any angle at all without appearances giving it away.

When we say an object is symmetric, we specify—or imply—the action that leaves it unchanged. The collection of all such actions comprises the symmetry group of the object. The symmetry group of an equilateral triangle thus has three members, that of a square has four, and more generally speaking, the symmetry group of the n-sided polygon has n members. These groups all have a finite number of members, reflecting the discrete nature of the corresponding symmetries. Such groups are interesting and useful, and they generate very pretty patterns, but for our present purposes it is *continuous* symmetries that are far more interesting, for these underlie gauge theories, such as quantum electromagnetism and the theories of the strong and weak nuclear forces.

The gauge theory formalism is immensely powerful because it not only predicts the action of a force, but goes deeper to explain why a force has the strength and range it does, and also why it mediates between particular particles; it does all this by drawing on arguments of symmetry. A continuous symmetry, as the name implies, is one in which the transformation (rotation, in the above examples) can be smoothly varied and is not restricted to discrete values. A simple example is the group of rotations of a circle. The angle can be varied uninterruptedly, and stopped at any point; throughout the process, the circle is left unchanged.

Spatial symmetries are familiar to us and hence are easy to recognize, but they are far from being the only kind; there also exist "internal"* symmetries—so called because they become manifest only in an abstract space, not in the visible three dimensions. These symmetries show up in a theory in the guise

---

*Such is the case for the gauge theories of electromagnetism, as well as the strong and weak nuclear forces.

of mathematical transformations that leave the equations unchanged.

As a basic example, consider the case where a quantity x figures in the equations strictly through its square x^2—there is no expression that contains x alone. Since $(-x)^2 = x^2$, we conclude that the theory is blind to the difference between x and $-x$. In other words, it is symmetric under the transformation of x into $-x$, and the theory is said to have a (discrete) symmetry. No part of this argument depends on the physical quantity represented by x; it could be anything at all. If, instead, the equations have such a form that they remain unchanged as x cycles through a whole range of connected values, the theory is said to have a continuous symmetry.

In order to give rise to a gauge theory, the underlying symmetries must be not only continuous but also local rather than global. What are global and local transformations, you ask? If the same transformation is performed throughout a system, it is said to be global; if the transformation can be varied across different points, it is said to be local.

As an example, think of daylight saving time. One fine day, an entire country wakes up and decides to set its clocks an hour forward. What was, until yesterday, 7 a.m. instantly becomes 8 a.m., but life goes on undisturbed and people can make and keep appointments the same as always, because the decision is a global one. Regardless of whether we call it 7 a.m. or 8 a.m., as long as the entire country has set its clocks forward by the required hour, everyone agrees on the time, which is all that is needed for the system of life to maintain an unchanged appearance.

A local transformation is far more interesting, because it has the potential to be far more chaotic. Suppose the people rise in protest, arguing for the right to determine their own individual times. They see no reason why the government, or indeed any agency, should tell them how to set their personal clocks. Eventually the political powers tire of arguing and give in. Since all citizens are free to set their clocks as they choose,* some precautionary measures clearly need to be taken in order to avoid widespread confusion.

Before we can decide what these measures should be, think for a moment about all that would remain unchanged. Irrespective of how we set our clocks, the duration of events would not be affected. School would still run for six hours, regardless of the fact that one child might say it lasted from 9 p.m. till 3 a.m., while his classmate would claim it started at 4:34 a.m. and ran until 10:34 a.m. People would be required to put in as many hours at the office as they did before; it would take the same amount of time to commute from one place to another or to cook a meal, and the same number of days would elapse between successive birthdays. In short, everyone would agree on how long something took, just not on the time at which it started or ended.

In the absence of a universal time, people would need to share their personal times with each other whenever they made appointments. If my dentist asked me to come in at 5:32 a.m., I would have no idea what she meant unless I asked what her current personal time was, and then tallied that against my own. This exchange of information would allow me to convert her time into mine, so that I could arrive at her office when she expected me. Such determinations of time are continuous (assuming the townspeople's watches can be varied in the infinity of subdivisions between seconds) and local (in that they change freely from person to person or place to place); thus they constitute gauge transformations.

It now becomes obvious that if a system (like a city) is to be symmetric (remain apparently unchanged) with respect to a

---

*Time, as it is viewed on a conventional clock, isn't quite a continuous symmetry, since traditional clocks show time in discrete seconds; think of this as an advanced civilization where time can be measured to infinite precision. People are no longer limited to saying "it is 9:45"; they can as easily declare it to be "8:35 and 49.234234984938 seconds."

ONLY THE LONGEST THREADS

gauge transformation—that is, while a particular parameter undergoes continuous changes that vary from place to place (like the determination of a personal time)—a transfer of information must take place (there must be a way to compare the time displayed on two distinct clocks). The mechanism for the exchange of information is what we call a force. The presence of this force is what makes the symmetry possible.

Physicists put it the other way around. Symmetry under gauge transformations gives rise to a force, they say. The very fact that system is invariant under a particular set of gauge transformations *implies* that an interaction exists to make this symmetry possible; gauge theories, therefore, come with interactions built-in. So, it is only natural to expect that the fundamental forces of nature should be described by gauge theories. And, indeed, they are.*

We have spoken of symmetries and transformations, discrete and continuous, local and global, but I have not yet said what the word *gauge* refers to. It is a tricky matter, knowing when to probe into the genealogy of words. Often names are just names, like quarks for instance—that word makes no sense. At other times, it is downright treacherous to read too much into names; the Up quark isn't any more vertically upright than the Down quark, and the Strange and Charm quarks aren't stranger or more charming than the Top or Bottom quarks. However, it just so happens that the use of the word *gauge* is somewhat revealing, and since gauge theories are an abstract concept, quite difficult to wrap one's head around, I thought I would tell you where the name comes from.

---

*It is somewhat of a moot point, almost an issue of semantics, whether or not general relativity is a gauge theory. The other three forces (electromagnetism, and the strong and weak force) can all be formulated as Yang-Mills theories, which means the gauge group satisfies a particular set of criteria (is a compact, semi-simple Lie group). Gravity stands apart because, even though it is a gauge theory in a broader sense of the word, it is definitely not of the Yang-Mills type.

The Greatest Forms of the Beautiful

In English, the word *gauge* refers to a measuring device, and this meaning carries over into physics. A gauge theory is one in which the means of measurement does not need to be universally agreed upon, but rather can be independently determined at each point in space. In other words, there is, everywhere, the freedom to "pick a gauge." The key point, of course, is that the physical world is blind to this choice; no matter how arbitrarily the measuring system is chosen from point to point, all appearances remain unchanged. A gauge theory contains concealed freedoms; there are a host of invisible choices we can make, without altering reality. But this liberty comes at the cost of introducing forces into the theory.

This intimate intertwining of forces and symmetries has spectacular implications; some of them are too technical for me to even attempt to share with you, but there is one which, given my current setting, is too tempting to resist. I turn to the tale of electroweak unification—the seminal work for which Abdus Salam, together with the American physicists Sheldon Glashow and Steven Weinberg, was awarded the Nobel Prize just two years ago.

The Nobel Prize is more than an award for excellence; it is a nod of acceptance by the physics community and signals the inclusion of a theory in the standard core of knowledge. The electroweak theory was subjected to some very rigorous testing before this prize was awarded. To my mind, at least, there is another argument in its favor: the fact that Glashow, Salam, and Weinberg, with their markedly different personalities, beliefs, and motivations, assembled the theory while working, for the most part, independently of each other.

Salam's motivations were almost religious. He credits his culture and heritage with having inculcated in him an appreciation for symmetry, a belief in unity, and the faith that a single underlying cause ties together disparate phenomena. Weinberg, on the other hand, considers religion "an insult to human dignity." In my mind, this dichotomy only renders their theory that much more beautiful. If two people with diverging beliefs can arrive at the same structure, surely it cannot be just a subjective construction of someone's mind; it must be a reflection of some objective reality that exists "out there."

Glashow, Salam, and Weinberg showed that the two apparently unrelated symmetries underlying electromagnetism and the weak nuclear force could in fact be traced back to a single, larger symmetry that had been "broken."

A quick comparison between electromagnetism and the weak force reveals the magnitude of this discovery. The contrast between the two forces couldn't be any starker! Electromagnetism is felt across large distances, but the weak force dies out at nuclear scales; electromagnetism is not chiral, but the weak force is; when subjected to electromagnetic interactions, particles retain their identity, but the effects of the weak force are alchemical and can transform one particle into another. Yet Glashow, Salam, and Weinberg managed to stitch these overtly different entities into one. Not since Maxwell amalgamated electricity and magnetism had physics experienced such a triumphant confluence.

Of course, there were many subtleties involved. The beauty and strength of gauge theories come from their heritage of symmetry, and this underlying structure had to be preserved. The key was to realize that symmetry was demanded only from the equations, and not from their solutions.

The idea that electromagnetism and the weak force might be related dates back to Julian Schwinger. Sheldon Glashow, his student, tried to make the connection, and a continent away, so did Abdus Salam and John Ward. Their theories were uncannily parallel, and both came up against the same problem: the weak force had to be mediated by massive bosons, but there seemed to be no way to endow force carriers with mass without blowing up the theory, so the work was put aside. A few years later, Abdus Salam and Steven Weinberg heard something that fired their imaginations. Drawing on recent developments in the theory of superconductivity, scientists had begun to question one of their most basic assumptions—the emptiness of the vacuum.

The vacuum isn't necessarily a void, they said, but merely the ground state, the lowest energy configuration—the baseline on which everything else is built, and from which nothing further can be extracted. It is possible, then, for the vacuum to be permeated by a field. Using this insight, Salam and Weinberg each arrived at the same solution: that bosons acquire mass through interactions with a new field that could be included in the particle physics repertoire without upsetting the symmetry of the equations. This Higgs field, as it came to be called, represented neither matter nor force carriers, but instead an entirely different—and previously unobserved—kind of particle known as a scalar. When the modified equations were solved, it was found that all the fields vanished in the vacuum, except the Higgs. The usual example given is that of a marble and a Mexican hat. Assume that the marble represents the state of the universe, the hat represents energy, and the value of the Higgs field corresponds to the distance of the marble from the center of the hat.

When the marble is balanced perfectly, right in the center of the hat's hump, the Higgs field vanishes, but the system has a finite energy, given by the height of the hump. The energy is lowered if the marble falls, and the vacuum corresponds to the lowest part of the brim (before the edges turn up again). When the marble falls to this position, the value of the Higgs field (the distance from the center of the tip) becomes equal to the radius of the hump. So, at one time, either the Higgs field or the energy can be minimized, but not both simultaneously. Poised on the hump of the sombrero, the marble is surrounded with infinite possibilities, each as good as the next, but its position is precarious and practically impossible to maintain. The energy of the system isn't minimized until the marble rolls down the hump, breaking this impasse. The direction in which the marble falls is completely random; the point on the rim where it lands is not distinguished in any way until, by virtue of the marble's arrival, it becomes the chosen vacuum, the point of reference for every-thing that happens from then on.

The details of electroweak theory are subtle, intricate, and almost breathtaking, and it can only really be done justice using mathematics. But a flavor of the argument is conveyed by the analogy Abdus Salam used in later years: "Look at ice and water," he writes. "They can co-exist at zero degrees centigrade, although they are very distinct with different properties. How-ever, if you increase the temperature you find that they repre-sent the same fundamental reality, the same fluid. Similarly, we thought that if you could conceive of a Universe, which was very, very hot . . . the weak nuclear force would exhibit the same long-range character as the electromagnetic force. You would then see the unification of these two forces perfectly clearly." Since we experience these forces only as they are now and not as they used to be in millennia past, we see electromagnetism as being inherently different from the weak force, whereas in fact the two can be traced back to the same root.

Once electromagnetism and the weak force had been sub-sumed into a single formalism, it was only natural to ask whether the strong nuclear force could also be incorporated. Abdus Salam says this is like asking if, in our analogy of ice and water, we can include steam. Several scientists began to won-der if the three quantum field theories could be encompassed in what came to be called a grand unified theory.* According to grand unification, the early universe was like an ideal and end-less water bath where the temperature was exactly the same ev-erywhere. Consequently, there was no means of distinguishing any one point from another, and every direction was exactly the same, from here on out to infinity. While these intensely high energies reigned and absolute symmetry prevailed, no struc-

*One such scheme was proposed by Salam and Jogesh Pati, another by Howard Georgi in collaboration with Glashow, and later, with Steven Weinberg and Helen Quinn.

ture was possible. As the universe expanded and cooled down, asymmetries developed naturally; the figurative water turned to ice. With its crystalline structure, ice distinguishes among different directions, whereas for water (at least under the idealized hypothetical circumstances we are considering) all directions are completely equivalent and interchangeable. Water is the more symmetric state, but ice has more structure: this appears to be a common theme.

It is by breaking the perfect symmetry of the early universe that the fundamental forces became manifest, and matter coalesced to form the elementary particles that congealed into atoms which dance around as the forces dictate, giving rise to our vast and varied world. Perfect symmetry might be aesthetically appealing, but in practice it is sterile. Figuratively, I suppose, one could say that the choice is between remaining frozen in stagnant perfection and descending into the frantic whirl of being; the complexity of life is possible only when things and places are different from each other and change is possible.

Craftsmen of old knew this truth instinctively. It is said that when a particularly enchanting piece of art was made, in which pattern, repetition, and symmetry played a role in enhancing the beauty and quality of the design, artisans would introduce a small but deliberate flaw on the grounds that the realm of perfect symmetry is reserved for the divine.

But, insipid as unbroken symmetry might be, a complete lack of symmetry would be far worse. Luckily, a benevolent providence has ordained for us the best possible combination. Sandwiched between the possibilities of stagnation and chaos, we find ourselves in this "best of all possible worlds," where there is simplicity of design but not implementation. The fundamental laws of our world may still be found by appealing to symmetry, but the applications are not thus limited. Even though the structure of our world is immeasurably rich, it is still economical in essence.

I know that this letter spans many new ideas, from the specific properties of a particular particle to general organizing principles in physics. Do not be frustrated if you can't grasp them all right now. Draw comfort from the fact that I have struggled with the expression of these concepts as much as, if not more than, you will struggle with their comprehension.

I was wrestling with words last night, striving for a clarity on paper that was perhaps not even there in my mind, when my eye fell on Rilke's *Letters to a Young Poet*. I bought this book in town last weekend, after learning that Rilke had spent some time in this area. He lived at Duino Castle, which, much like Miramare, is perched on a cliff over the cerulean waters of the Adriatic. There is a walking trail, just over a mile long, that leads up from the fishing village of Sistiana to Duino. It is abandoned now, but since it is named after Rilke, I like to imagine that it is a path he trod often, perhaps while pondering life's great mysteries.

I had started reading the *Letters* but had not yet gone very far. Every paragraph demanded to be savored, and I was reading slowly to extend my pleasure. When I picked up the book last night, I came upon a passage that spoke directly to me. A gentle admonition floated up across the years, and it was exactly what I needed to hear. I cannot think of any more fitting way to end this letter than to share it with you.

Rilke wrote, "Have patience toward all that is unsolved in your heart and try to love the questions themselves, like locked rooms and like books that are now written in a very foreign tongue. Do not now seek the answers, which cannot be given you because you would not be able to live them. And the point is to live everything. Live the questions now. Perhaps you will then gradually, without noticing it, live along some distant day into the answer."

I hope and pray that both of you learn to delight in walking with questions, and that you are granted the exultation of arriving at answers. The very wise Henri Poincaré wrote, "The scientist does not study nature because it is useful to do so. He studies it because he takes pleasure in it, and he takes pleasure in it because it is beautiful. If nature were not beautiful, it would not be worth knowing and life would not be worth living."

I wish for you to truly experience—and revel in—this "intimate beauty which comes from the harmonious order of [nature's] parts." Of earthly joys, not many are greater than this, except of course love—and that you will always have, for to you belongs my whole heart.

Your adoring Abba

P.S. Hassan, I purposely added more postage than was needed, on both this letter and the one I wrote your mother. As you must have noticed, each stamp is different. You may keep all of these for your collection, but please do not drown your sister's postcards in an attempt to pry the stamps off—I have used only duplicates on those.

CHAPTER 6

Mystery Has Its Own Mysteries

[Quantum Field Theory and Renormalization]

*Mystery has its own mysteries, and there are gods
above gods. We have ours, they have theirs. That is
what's known as infinity.* —Jean Cocteau

8 December 1999
Stockholm, Sweden

Assorted winter hats of different shapes and colors trace an unbroken path from the main door of Aula Magna all the way
down Frescativägen, finally disappearing into the underground
warmth of the Universitet subway station. Some of the hat wear-
ers, done with the day's festivities, are going back home or even
to work. Others are heading for a quick warm lunch in the hour
or so that remains before the economics laureate takes the stage.
I don't intend to return, but I am in no rush to leave. This morning's Nobel lectures have left me somewhat giddy. My thoughts
are heady, exhilarating even, but my limbs are strangely lan-
guid; the conflict within me could not have found a more literal
representation.

It has been a while since the physics award was made for
a theoretical discovery. I came here this morning to relive the
euphoria of being swept up in a tide of thrilling ideas, not to
revisit the raft I had abandoned by the side. But somewhere deep
within, I must have known what this combination of excitement
and nostalgia would lead to.

A band of blond Scandinavian wood runs across the huge panes of glass that curve around me. December's golden rays pour in, tinting the atrium with a gentle glow. The sky is incandescent, glimmering with the promise of snow. In this mellow light, emotions I had long pushed back return to cast their dark shadows upon my mind. I feel increasingly irritated and impatient. I want to shake myself. Perhaps it is time to bring those neglected demons out in the open and deal with them once and for all, and this light and airy building might be a good place to start.

With a sigh of resignation, I step out of the animated queue heading toward the door. I let the crowds slide by, like snow slipping off a roof, until the chatter ceases and I stand almost alone in the foyer. A cold haze filters the view, blurring the centuries-old oaks around Aula Magna and smoothing over the angular edges of nearby buildings. In its quiet reverence for winter, the world is a picture softly focused.

I linger for a while by the large windows, watching, procrastinating, until the sharp sound of a closing door brings me out of my reverie. Maybe I should do this quickly, the way you rip off a band-aid. The questions that haunt my days are these: Did I give up on academia too easily? Did I "settle" when I chose a job in industry? Can you really claim to have loved something if you can turn around and never look at it again? All those years of graduate school, all those intense highs and lows, that obsession with physics, passionate as any love affair, was all that even real? Did I deserve to continue? Maybe the awful truth is that I did love physics, but it never loved me back.

Even now, with all the years that have passed, that thought makes my eyes sting. I need to think of something else. Reflexes take over, and I turn instinctively to my oldest coping mechanism: treat the mind like a toddler. Distract it with idea toys.

And right now, there's one very obvious thing to play with: the fact that Gerard 't Hooft and Martinus Veltman are being awarded this year's physics prize for "elucidating the nature of the electroweak interaction." A remarkably short epigram for the long and arduous journey that began, about three-quarters of a century ago, with the confluence of quantum mechanics and special relativity. During his lecture this morning, Veltman mentioned the many great scientists who had paved the way for him and 't Hooft; behind almost every sentence he uttered stood a Nobel laureate, armed with decades of work, backed by a staggering army of soldiers—most of whose names are lost in the annals of history.

Toward the end of the 1920s, quantum mechanics had many feathers in its cap, but despite these successes the theory was riddled with omissions and—even worse—apparent contradictions. One problem was the phenomenon of particle creation and annihilation. Prior to special relativity, the number of particles involved in any process was thought to be a fixed, immutable quantity. When he wrote down the most famous equation in history, declaring mass to be just another form of energy, Einstein unleashed the disturbing possibility that energy could manifest itself as particles that appeared seemingly from nowhere—like rabbits popping out of a magician's hat—and later could disappear, just as easily, into the void. There was no provision in quantum mechanics for this unnerving behavior, and it became clear that the theory had to be modified to allow for particles that flickered into and out of existence.

Another thorny issue had, however, been more or less resolved. Physicists learned to embrace the wave-particle duality that had once driven them to despair. It turned out that the conflict had never been one of fact but of interpretation. In the pre-quantum days, we had a pretty decent understanding of both particle and wave behavior. We had various sets of equations, which told us how a ball would fall under the influence of gravity, how a charged particle would move in an electric field, or how a sound wave would disperse in a particular medium. The mathematical form of the equations describing the behavior of waves were quite distinct from those for particles, and this difference was reflected in the vastly different ways particles and waves conducted themselves.

Eventually, and inevitably, we began equating the behavior with the equation, so that if a particular system evolved in a manner that could be described by the mathematics of the wave equation, we referred to it as a wave, even if there was no physical wave anywhere to be seen; the concept had become abstract, and the "wave" or "particle" label was applied more to the mathematics describing a system than to its visible attributes.

A new paradigm was needed, and this came with the acceptance of the field as a fundamental physical entity in itself. Even though, as Einstein wrote, the concept of a field had entered physics as "no more than a means of facilitating the understanding of phenomena," over the years it became, for the modern physicist, "as real as the chair on which he sits." In order to be compatible with quantum theory, fields, like everything else, had to exist in discrete multiples of a fundamental unit; what we had previously thought of as particles were really just these little "clumps" of the field, whereas waves represented the probability of finding a particle at a particular location.

I still remember when this idea really clicked in my mind. I was flying back home after a weekend trip to Gothenburg. It was dark, and as we broke through the cloud cover, I saw the city mapped out in light. From that high up, I couldn't make out individual flickers; a wash of brightness lay over Stockholm, thicker in some places than in others, outlined sharply by the black waters of the archipelago. As we continued our descent, the lights grew larger and more distinct; each began to separate itself from the crowd, and then it hit me—this was a pretty good illustration of a field! When viewed from a substantial distance, a large distribution (of particles, or lights) appears to be continuous and can be described quite naturally in terms of intensity; dark, empty spaces correspond to zero intensity, whereas crowded, well-lit areas represent high intensity.

ONLY THE LONGEST THREADS

But what if we want to talk of an individual object rather than a whole ensemble? Say I took an aerial photograph when a million lights were shining in the city. I don't expect to see an identical view, only a million times dimmer, when a solitary bulb lights up the dark. My intuition tells me that the vocabulary I used for a continuum cannot be applied to a single entity—but mathematics says otherwise. The equations for one speck have the same form as the equations for a million specks; when we sketch them out as graphs, we get identical shapes. The only difference is in the physical interpretations. In the case of a single bulb, the light map of the city should be interpreted as a measure of *probability*: it is more likely that the bulb be lit in the brighter areas of the map than in the darker areas. What was previously a literal image of the intensity of light across space now represents the chance that our bulb is lit at a particular location.

With this leap in understanding, wave-particle duality ceased to be a dichotomy; both these apparently contradictory facets of reality were subsumed in the field formulation. It became clear that, at the quantum level, waves describe probabilities, but when the number of quanta is large (i.e., in the classical limit), probabilities begin to look quite like certainties, and waves can be said to represent the strength of the field.

As this new paradigm was applied to the electromagnetic field, the field quantum (dubbed the photon) was interpreted as a particle that carried a discrete parcel of energy and momentum and traveled at the speed of light. The attraction (or repulsion) between two charges could now be attributed to the exchange of photons between them.

The conflict between special relativity and quantum mechanics was finally resolved at the end of the 1920s when Paul Dirac brokered a brilliant peace settlement. Stripping the problem to its essentials, Dirac considered the simplified case of an electron-only universe. The only relevant force in this isolated reality is electromagnetism: electrons are light enough that the effects of gravity are practically negligible; being leptons, they are immune to the strong force, and the weak force isn't an issue either because there is no other form of matter for the electron to transform into. Thus, the sole interaction is due to the electric charge on the electrons. Since, according to quantum theory, electromagnetism is mediated by photons, Dirac's toy universe automatically contained photons in addition to electrons. An electron generated an electromagnetic field by emitting photons, and responded to the field by absorbing photons.

Dirac also made allowances for special relativity by incorporating the possibility of photons being spontaneously created (from energy), and annihilated (into energy). Thus incorporating the dictates of quantum mechanics and relativity, Dirac worked out the mathematics that governed the behavior of an electron. The resulting equation was an absolute triumph. Not only did it replicate known results, it was also prescient: the spin of the electron—that troublesome fourth quantum number, postulated *ad hoc* just to explain experimental data—arose as a natural and inevitable consequence.

There was, however, a tradeoff, which caused Dirac much concern. His equation did not make a single assertion, as most equations do. Instead, it gave rise to a set of *four* interrelated statements, each of which had an associated mathematical solution. Two of these solutions were much lauded because they described the electron in each of its two spin states, but the remaining two solutions were rather embarrassing; they seemed to indicate the existence of a previously unsuspected particle, identical to the electron in every way, but carrying the opposite charge. Dirac struggled with this for a long time, and tried unsuccessfully to eliminate this seemingly ridiculous possibility, but there seemed to be no consistent, sensible way to rid the theory of this nuisance. It soon became clear that the equation knew what it was saying: the anti-electron, also known as the positron, was seen in experiments shortly thereafter. Mathematics had somehow "known" about the existence of this particle even when we had no physical reason to suspect its presence.

The elegance of Dirac's equation was unparalleled, and the secret knowledge woven into the mathematics quite uncanny, but it had one major drawback. Dirac allowed for the photon to be a relativistic object, flitting in and out of being, but he had not given the electron similar leeway; until that possibility was incorporated, we could not claim to have a theory fully compatible with special relativity.

But Dirac *had* managed to create a mechanism whereby the creation and annihilation of particles could be systematically wired into a quantum theory. This method would be used countless times in the decades to come, as relativistic theories were formulated to describe electromagnetism and the strong and weak forces at the quantum level: these came to be known, not surprisingly, as quantum field theories.* The quantum field theories for the weak and strong forces were more complicated than those for electromagnetism, mostly because they had more force carriers, which led to more complicated interactions. The eight force-carrying quanta, or gauge bosons, of the strong force are called gluons, whereas the weak force is mediated by a group of three gauge bosons, somewhat prosaically called the W^+, W^-, and Z. In keeping with the dictates of special relativity, these force particles can materialize and disintegrate at will (whenever the corresponding field has sufficient energy), just as the matter particles—the quarks and leptons—can.

The first quantum field theory to be given coherent form was quantum electrodynamics. QED, as it was called, was a natural extension of Dirac's equation, with the caveat that both electrons and photons now had an equal right to materialize out of,

---

*Or, more properly, relativistic quantum field theories, but for the most part they are referred to simply as quantum field theories.

and disintegrate back into, the void. One could have been pardoned for thinking that, once we had this appearing/disappearing mechanism down for photons, including electrons would be just a corollary—but that turned out to be far from the truth. Regarding photons as transient states of energy was one thing, but considering electrons to be similarly ephemeral came with a whole other set of challenges.

This was due mainly to the following fact: Dirac's equation described only two basic processes—the emission, and absorption, of a photon by an electron. In quantum electrodynamics, another kind of interaction became possible: an electron-positron pair could be created from, or dissolve into, a photon. This led to an unending series of ever more elaborate gymnastics. A photon could disintegrate into an electron-positron pair, of which the electron, in turn, gave off another photon, which went ahead and did the same; since an infinite number of maneuvers now became possible, the theory quickly spiraled out of control.

As Heisenberg put it, the discovery of antiparticles changed everything. "As soon as one knows that one can create pairs," he said, "one has to consider an elementary particle as a compound system; because virtually it could be this particle plus a pair of particles plus two pairs and so on, and so all of a sudden the whole idea of elementary particles has changed." Elementary particles, hitherto considered to be the simplest possible objects, suddenly became complex, constantly evolving, dynamic systems.

When I first learned about this a decade ago in graduate school, I found the domino effect fascinating, but I couldn't see why it was such a huge catastrophe. After all, there had to be some logical limit to how far the chips could fall, right? Wouldn't the constraints of energy conservation rein in the madness?

Apparently not. I was simultaneously scandalized and spellbound when I found out that even sacrosanct laws have (quantum-sized) loopholes, thanks to the uncertainty principle. The argument goes thus: There is an inherent ambiguity in simultaneous measurements of energy* and time, and the product of these uncertainties must be larger than Planck's constant. The more precise one measurement becomes, the fuzzier the other. In Eddington's vivid language, "An addition to knowledge is won at the expense of an addition to ignorance. It is hard to empty the well of Truth with a leaky bucket." Over sufficiently short durations, our measurement of energy is so vague that even massive fluctuations would remain unnoticed. As a result, huge amounts of energy can be pilfered from the universe by thieves who act too quickly to be caught.

The stolen energy can manifest itself as a particle and anti-particle pair that flickers in and out of existence so fast as to escape detection. In sharp contrast to the law-abiding particles with which we are familiar, these "virtual" particles are subject to no constraints. They may live on borrowed time, but during their brief existence they are free do what they please. They may even break the sacred speed barrier set by light, because they can never go down on record. Their behavior can flout all expectations; it does not even need to make sense.

At first glance, these delinquent phantoms appear too ridiculous to be taken seriously, but they turn out to be essential; they are responsible for mediating the forces. Consider, for instance, the mechanism of repulsion between two electrons. In quantum field theory, we attribute this to an exchange of photons. One electron emits a photon (thus generating a field) and the other electron absorbs said photon (thereby responding to the field). If energy is to be conserved, the electron that emits a photon should lose energy and hence recoil, while the electron absorbing the photon should gain energy and hence be pushed forward. This picture explains, quite satisfactorily, the repulsive motion of electrons, but it breaks down when the electrons stop

---

* An analogous statement holds for momentum and position. The values of both cannot be simultaneously determined to an arbitrary degree of accuracy.

moving. Even when they are stationary, the repulsive force between electrons continues to exist, and it follows that photons must still be exchanged. But it flies in the face of energy conservation that this loss and gain of energy do not affect the electrons' motion. What's more, a detector placed in between the electrons records no such passing photons. This lack of a trail points, almost conclusively, to virtual photons. This sounds like a desperate, though elaborate, argument, but the proof is in the mathematics. Under the assumption that virtual photons are exchanged between electrons, the equations of QED are able to reproduce, exactly, the magnitude of the repulsive force.

It is mind-boggling but true; in order to reproduce the familiar in the visible parts of our theory, we must make allowances for all sorts of preposterous happenings in the dark crevices that remain hidden from view. Faced with these arguments, we grudgingly accepted virtual particles and rearranged our understanding to make room for them. But just as soon as we made that concession, the flip side of our decision became apparent, and its consequences blew up in our faces.

Unlike classical mechanics, which gives a single answer to every question, quantum mechanics comes up with a list of possible outcomes, starting with the likeliest. In working out the answer, you have to take into account every conceivable way in which the process could have taken place. Say you want to calculate the path an electron took in going from one point to another. In classical mechanics, you would draw a single curve and be done. In quantum mechanics, you have to account for each path that connects the beginning and end points, whether or not it is a plausible option. Having opened Pandora's box of virtual particles, we must now include even blatantly unphysical paths for virtual particles to wander down. Unconstrained by reason, these paths can always be embroidered further: a virtual electron-positron pair can be inserted into the path of every photon; before they decay, the interim electron and positron can, in turn, emit a photon, which can again be replaced by another electron-positron pair, and so on and so forth, ad infinitum.

In this madness, we clung to the hope that, as often happens in mathematics,* an infinite series would somehow yield a finite sum. At first glance it seemed things might work. Even though all sorts of elaborate routes had to be included in the calculations, each added flourish made them increasingly unlikely. Much to our relief, the contribution of every successive term was smaller than that of the one before it. It seemed the series might converge, but when we added it all, the answer blew up anyway.

The situation was as bleak as a Scandinavian December, and a glimmer of hope was sorely needed. When the days shrivel up, in preparation for the winter solstice, in every town across Sweden, processions of little girls, following in Saint Lucia's wake, light up the dark. With voices sweetly harmonizing, the girls sing of brooding shadows and of Night who walks with her hushed, heavy feet, until Saint Lucia appears to dispel the gloom. Clad in white, a wreath of candles in her hair, Lucia whispers of the rosy dawn to come. Fortified by this vision, and aided by glögg, gingerbread cookies, and raisin saffron buns, we get by until the nights grow shorter and the gorgeous sun has his day.

No haloed angel appeared to guide QED out of the dark; that journey was less poetic, but the relief was equally profound. Many brilliant scientists grappled with the problem of divergence, as it is called—these infinities that seemed to bounce off the walls, growing ever louder, adding layers of echoes to the cacophony. It took a group to quiet things down, but if there is one name associated with this effort, it is that of Richard Feynman; he was the man who brought QED to the masses. Feynman, the

---

*For example, you can generate an infinite series simply by writing down a string of numbers, each of which is half the previous one. If you add all these terms together, you will get twice the number you started with. For example, $1 + \frac{1}{2} + \frac{1}{4} + \frac{1}{8} + \frac{1}{16} + \frac{1}{32} + \ldots = 2$.

larger-than-life, safe-cracking, bongo-drum-playing showman and trickster, was awarded the Nobel Prize along with Julian Schwinger and Sin-Itiro Tomonaga for delving into the heart of infinity and coming out with a finite, sensible answer. These three scientists conducted their work independently, and their approaches to the subject might as well be character sketches.

Even though the world learned of his work later, Sin-Itiro Tomonaga came first. During the enforced seclusion of World War II, Tomonaga worked on "renormalization"—the attempt to deal with the unwieldy infinities—unaware of the experimental results that nudged and prodded Feynman and Schwinger on their way. Yet, within that perfect silence, drawing on theoretical considerations alone, this gentle Japanese man came to the same conclusions that the other two would also eventually reach.

Schwinger was a child prodigy. He worked on advanced research problems while still in high school, and became a Harvard professor at age 29. An immaculately dressed classical music connoisseur who drove a shiny black Cadillac, Schwinger worked in the rigorous mathematical tradition. His formulations were like textbook French gardens, meticulously manicured and perfectly watered, each leaf and flower in place. Lesser mortals could marvel at his expertise, but they could not reproduce what he had done.

Feynman, on the other hand, was brash, iconoclastic, and almost deliberately unsophisticated. His physics was intuitive. He consigned tradition to the winds and constructed his own explanations. Feynman made a virtue of taking physical phenomena apart and putting them back together without the instruction manual. These traits served him particularly well when it came to navigating the fractal labyrinth of quantum electrodynamics.

Unencumbered with the need to use conventional tools, Feynman came up with a system of simple diagrams that he used to represent, and track, the contributions from various physical processes as he proceeded through line upon line of calculation. Armed with an incisive curiosity and what appeared to be a collection of irreverent cartoons, Feynman succeeded in chasing the infinity down to its source. Surprisingly enough, the dissonance arose from only three places. Immediately, the problem dwindled to a manageable size. If only we could tame three divergences, all the noise would melt away and we would finally hear the wonderful melodies we were after.

Feynman's pictorial formulation was easy and intuitive, whereas Schwinger's was formal and complex. These two perspectives clashed, just as Heisenberg's matrices and Schrödinger's waves had done a few decades earlier. Once again, a dictionary had to be established in order to prove that the statements made by the two competing formalisms were, in fact, the same. As Dirac had performed this task for quantum mechanics, Freeman Dyson did it for QED.

The story is charmingly told in a modern-day fable that forms the preamble for a series of lectures on quantum field theory delivered a decade and a half ago by Predrag Cvitanović at Bohr's institute in Copenhagen. I had a photocopy of the notes, and I pinned the introduction onto the wall just above my desk. I looked at it every day for years. I could probably recite it verbatim even now.

Once (and it was not yesterday) there lived a very young mole and a very young crow who, having heard of the fabulous land called Quefithe, decided to visit it. Before starting out, they went to the wise owl and asked what Quefithe was like.

Owl's description of Quefithe was quite confusing. He said that in Quefithe everything was both up and down. If you knew where you were, there was no way of knowing where you were going, and conversely, if you knew where you were going, there was no way of knowing where you were. The young mole and the young crow did not understand much, so they went to the eagle and asked him what Quefithe was like. The eagle shook his white-feathered head, sized them up with his fierce eyes, and said: "Action gives automatically invariant description of Quefithe. You must study the unitary representations of the Lorentz group." The mole and crow waited for more, but the eagle remained silent, his gaze fixed on an unfathomable string in the sky.

Clearly, if they were ever going to learn anything about Quefithe, they had to see it for themselves. And that is what they did.

After a few years had passed, the mole came back. He said that Quefithe consisted of lots of tunnels. One entered a hole and wandered through a maze, tunnels splitting and rejoining, until one found the next hole and got out. Quefithe sounded like a place only a mole would like, and nobody wanted to hear more about it.

Not much later the crow landed, flapping its wings and crowing excitedly. Quefithe was amazing, it said. The most beautiful landscape with high mountains, perilous passes and deep valleys. The valley floors were teeming with little moles who were scurrying down rutted paths. The crow sounded like he had taken too many bubble baths, and many who heard him shook their heads. The frogs kept on croaking "It is not rigorous, it is not rigorous!" The eagle said: "It is frightful nonsense. One must study the unitary representations of the Lorentz group." But there was something about the crow's enthusiasm that was infectious.

The most puzzling thing about it all was that the mole's description of Quefithe sounded nothing like the crow's description. Some even doubted that the mole and the crow had ever gotten to the mythical land. Only the fox, who was by nature very curious, kept running back and forth between the mole and the crow and asking questions, until he was sure that he understood them both. Nowadays, anybody can get to Quefithe—even snails.

I notice a hum of voices around me. People are beginning to trickle back in for the economics lecture. I really should head home. As I loop my scarf around my neck and wiggle my fingers into gloves, I catch, almost by accident, a smile playing at the corners of my mouth. The memory of this story has caused a rush of affection and warmth for Quefithe and what it represents, as well as for those who wrote the little fable, those who chuckled over it, and those who laid the foundation for a story to be told at all. Suddenly, I know with a deep certainty the answer to at least one of my troubled questions. What went on between me and physics was love. I loved it then, and I love it still. When they are not pelted with insecurity, my heart and mind respond to physics instinctively, and with delightful confidence.

The subway platform is full of students, as always. As I wait for my train, I reacquaint myself with the mosaic of knowledge along these walls: countless figures, diagrams, words, and symbols, painted on little tiles, hinting at all that can be learned when you alight at this stop. Each of Stockholm's Tunnelbana stations is decorated in a unique manner. There are many I like, but Universitet has always been one of my favorites.

I have made this trip so many times in years past that muscle memory kicks in. Without being aware of how it happened, I find myself seated on the train. That is how it often happened, all those years ago. This business of being lost in thought is familiar too. We glide smoothly between stations, through the dim underground passage. Vivid reflections of the subway car flicker in the windows. I catch a glimmer of myself, and for a split second I see the girl I was when I last visited this campus: a knapsack-toting student, slouching under the weight of the library books I had come to return. Ten years went by in a flash. It feels like she should be here, on this train, maybe in a different car . . .

If I ran into her on the platform, what would I say? Sci-fi movies insist that you can't say anything about the future when you go back to the past, but I don't think I'd even want to. If I met my younger self, I'd talk to her about what's on my mind right now. Odds are, the same things would be on hers too. Things like renormalization.

It is a fascinating concept, and one that really should not work, but does anyway. If you told mathematicians you were trying to subtract infinity from infinity in an attempt to get a finite answer, they would either laugh you out of town or have you committed. Physicists, however, pit infinities against each other as a matter of routine; but we weren't always this debonair. It all started with the discovery that the infinities messing up our calculations stemmed from only three basic sources: the electron's mass, its charge, and vacuum polarization—the newly discovered ability of particle and antiparticle pairs to bubble up in the vacuum. Despite the fact that the mass and charge of the electron are quite obviously, and measurably, finite, the mathematical apparatus of quantum electrodynamics insisted otherwise. We wondered why the theory should respond so perversely to our line of questioning. Why would it hand us answers that were so blatantly untrue?

Indignation turned into chagrin when we discovered that the mathematics was not lying; rather, it was our questions that broke the rules. Despite repeated rebuffs and warnings to stay outside the quantum, we kept trying to penetrate the inner sanctum. Some of this was unintentional; many of our patterns of thought were still mired in the classical tradition, and we sometimes broke the quantum rules without meaning to. But, however heedless our trespass, it could not be forgiven. When we strayed close to the protected confines of the quantum, nature chased us away with expletives and profanities; she hurled infinities at us.

The verdicts of mathematics are absolute; they cannot be transformed into more desirable statements through mere argument or clever manipulation. In order to arrive at a different answer, we had to question the physics and discover where we had gone astray. The problem was found to stem from the old picture of particles as point charges. Points have no place in a quantum world; space simply cannot be parsed that fine anymore.

Our insistence on dealing with mathematical ideals was strangling physics.

Consider the electron. It carries charge and therefore an associated electric field; this, in turn, contains energy or, equivalently, mass. So the charge of an electron contributes to its mass. The quantity we measure in experiments is the sum of the electron's "bare" mass, i.e., the mass of a hypothetical, uncharged electron, together with the mass of the electric field. This composite mass is all nature grants us access to. In a similar vein, despite the cleverness of our experiments, we can never measure the electron's bare charge; it is shielded from us by the virtual particle pairs that bubble in and out of existence around it. Virtual positrons, attracted by the electron, swarm around it and screen its charge, whereas virtual electrons are repelled away. Since we are unable to penetrate the positron cloak, the quantity we measure is what remains of the electron's charge after the positrons have neutralized some of it. A parallel argument takes care of the infinities caused by vacuum polarization.

Such enhancements and reductions happen around any charged particle, regardless of its size, but the effects are more pronounced for smaller particles, whose associated fields are stronger and more focused. If particles were indeed of vanishing size, true "points," the fields surrounding them would be infinitely dense, as was reflected in our equations. Our presumptions were clearly wrong, but QED had too many successes to be scrapped entirely. In a certain realm of applicability, the theory yielded answers of extraordinary accuracy. And so we came to realize that it was not incorrect, merely limited. The finiteness of the electron's mass and charge had to be accepted as articles of faith. QED could not be used to derive the values of these quantities, but it could calculate everything else relative to them.

Conceptually, that makes sense. The way it was implemented, however, caused eyebrows to be raised. We were no better than students who, stuck on a particularly hard problem set, turn to the answers at the back of the book and then reverse-engineer the problem so that the desired answers emerge from the equations.

The only way to make this approach work was to concede that bare quantities were infinite, and claim that after the infinite effects of the field were accounted for, the "dressed" quantities would be finite. It was as if two unimaginably immense forces were playing tug-of-war. The forces themselves were beyond our scope of measurement, but the quantities they were fighting over moved just a little. In order to obtain sensible results, we had to content ourselves with measuring these movements and avoid making conjectures about the rival forces that tugged at the rope. For physicists, who are notoriously curious people, constantly peeping through "the keyhole of nature, trying to know what is going on" (as Jacques Cousteau put it), this was a very hard pill to swallow.

Even the usually reticent Dirac was moved to vocal criticism of this "arbitrary" way of discarding infinities. "This is just not sensible mathematics," he protested. "Sensible mathematics involves neglecting a quantity when it is small—not neglecting it just because it is infinitely great and you do not want it!" Extrovert Richard Feynman was as different from Paul Dirac as two people could be, and yet he was equally disturbed at having to resort to such "hocus-pocus," which he also suspected was not "mathematically legitimate." "This shell game that we play," he said, "is technically called 'renormalization.' But no matter how clever the word, it is still what I would call a dippy process!"

That might well be true, but a better formulation still hasn't been found. The fact remains that, while there is probably an inadequacy in our current formulation of quantum electrodynamics, it does provide a reasonably efficient machine for cranking out quantitative answers. For now, we use what we have. Of course, this machine isn't the most elegant thing to watch in motion. Calculations are done in a hierarchical manner: all interactions with the same number of virtual particles are placed on one tier, and an interim answer is computed; we then proceed, tier by tier, further refining our answer with each successive step. All our calculations, no matter how accurate, are eventually truncated. We can go only so far. An exact result, an analytic expression for the sum of the infinite series, does not yet exist.

When we study QED in graduate school, we learn to crank the machine by hand and perform these rather tedious calculations. I remember my frustration when theory turned into practice and I had to get in the ring with this tangled mess. Sorting out the various strains of infinity, aligning them so that their opposing pulls led to balanced, renormalized behavior, was far from being a trivial task. I was quite overwhelmed, in fact. One day as I stared at the papers littered around me, covered with scribbled equations, I thought of long-forgotten childhood stories about the Tomtar. If the legends are true, these capable little gnomes are sticklers for order and refuse to frequent messy places. The Tomtar bring luck and prosperity to their hosts. They crawl out at night, eat what has been left out for them, and get to work. By morning, all necessary chores have been done.

How wonderful would it be if a Tomten* could help me sort out this maddening jumble of infinities, I thought. I sprang into action and organized my office. I sorted out the books on my shelves, returned those I no longer needed to the library, cleared out the random old things that clogged my desk drawers, stacked my notes into neat piles, and cleaned up my desk. Next morning, I brought a ceramic Tomten figurine with me to school. A tall, pointed red hat covered his eyes, and a little round nose stuck out over the white beard that flowed down over his gray gown. During the day he lived in my top drawer, but every night, before I went home, I cleaned up my desk, leaving out only those calculations that troubled me, and I placed my little gnome next to them. My officemate thought this was insane until she witnessed the results. In the mornings, with Tomten

* *Tomten* is the singular of *Tomtar*.

back in his drawer, I hunched over my calculations again, and somehow they made more sense than they had the night before. On occasion, the emergency stash of candy in my office seemed unusually depleted, but my work began to fall into place.

That Tomten lived in my office until I graduated. He's probably crammed between books and notes in the box where I store my "physics stuff." I should look for him when I get home. There are still tangles in my life he could help with. I might need to pacify him first, apologize for the long years of neglect, ply him with sweet treats, but I think he'll eventually give in.

The train stops at Stadion. Amid the customary shuffle of footsteps, people get off and on. The old man in front of me unfolds a newspaper. As he ruffles through the pages to find his spot, I see a news item about the Nobel Prize. The winners are delivering their public lectures today, it says, but the award ceremony takes place the day after tomorrow on the anniversary of Alfred Nobel's death. The article lists the times, venues, and titles of the lectures. I hadn't noticed earlier, but now that I see it printed in black and white, I am struck by the fact that 't Hooft called his talk "A Confrontation with Infinity." Upon reading that phrase, the first image that comes to mind is Carl Milles's *The Hand of God*.

Several reproductions of this statue exist around the world. I do not know how they are displayed elsewhere, but the one I have seen is silhouetted against the sky on top of a long narrow column in the sculpture garden at Milles's house in Lidingö. It makes an arresting picture: a man poised, somewhat precariously, between the thumb and index finger of a large left hand floating in the sky. His tension is apparent. Legs bent at the knee, indicative of an uneasy balance; arms spread out, palms upward, in a gesture that is part confusion, part curiosity; head tilted back, almost too far, in his need to gaze up, presumably at the Creator who holds him so gingerly. Milles's sculpture is one powerful representation of a confrontation with infinity; the lecture this morning was another.

I knew 't Hooft had done his Nobel Prize–winning work in his mid-twenties, but I was still surprised at how young he looked. With a quiet, understated authority, he picked up the thread where his Ph.D. thesis advisor, Martinus Veltman, had left off. He recounted the history of renormalization, its success in QED, and its failure to deal with the weak interaction.

The problem with the weak nuclear force was that all three of its gauge bosons, the W^+, W^-, and Z, were massive. On one hand, this made a lot of sense; heavy force carriers can travel only small distances, which explains why the weak force has such a short range. On the other hand, it caused major issues. The existing techniques of renormalization applied only to theories with massless gauge bosons, like QED and its photon. The presence of a massive gauge boson led to an infinite number of divergences—an infinite number of quantities whose values must be experimentally measured—and a complete loss of predictive power. Physicists strained to extend the techniques of renormalization to include massive gauge bosons, but until Gerard 't Hooft, none succeeded.

As a young graduate student, 't Hooft showed that the problems could be circumvented and a theory be liberated from infinities, provided its gauge bosons acquired their mass *through an interaction*. According to electroweak theory, this is precisely what happened to the W^+, W^-, and Z. They all started out on an equal footing with the photon—massless. It was only the breaking of symmetry, and the subsequent presence of a Higgs field, which led to the four gauge bosons being apportioned different masses and electromagnetism being distinguished from the weak force.

The argument is rather interesting, and it hinges on the fact that the presence of an asymmetric field in the vacuum affects our perception of mass. This is not that hard to understand. Say two metal balls are placed on a flat surface, and you are asked to judge their weight without picking them up. An instinctive reaction would be to try pushing them. Experience, or intuition, tells us that if the same force is exerted on both, *regardless of which direction you push them in*, the heavier object will move more slowly, and the lighter one will go faster. If they move at the same speed, you would expect them to have the same mass.

Suppose that I now turn on an electric field in one particular direction; with that choice, I break a symmetry. Since the field is invisible, it appears that nothing has changed. But should you push the same two balls again with identical force, you might see one ball move more slowly than the other. You would conclude that this ball was more massive, whereas in fact the discrepancy might arise because this ball was charged and was pushed in a direction where it was repelled by the field and thus slowed down. In other words, particles that appear identical in the absence of a field can take on a variety of appearances in the field's presence.

Mass, which we had assumed to be an intrinsic characteristic, turns out to be a perceived quantity. It is a physical manifestation of the interaction between an object and a background field. In a similar vein, the W^+, W^-, and Z bosons acquire their masses through an interaction with the Higgs field; and because this is the case, 't Hooft's methods could be used to renormalize electroweak theory and extract sensible predictions from its equations.

With an uncharacteristic screech, the subway pulls into T-Centralen. I had intended to go straight home, but I have a sudden urge to see the sun set over the water, so I get off the train and head out. The smell of food assaults me at the Sergels Torg exit, and I realize I am starving. Perhaps a quick trip to the NK Konditori is in order. The lowest floor of Stockholm's ritziest department store is a gastronomic wonderland. I avoid coming here because I always surrender to temptation. Today, I don't even pretend to resist. I buy a bottle of cloudberry jam to take home, and for now, I indulge in the decadent chokladbiskvi: moist, crumbly almond cookies topped with whipped buttercream and coated in chocolate glaze.

Hunger quelled, my thoughts return to physics. The Higgs field is central to the story, yet the only proof we have for it thus far is our need for it. Every other particle in our description of the universe has made its presence—or at least its absence—felt. The Higgs boson stands alone in that there is no physical reason for us to assume it exists. Not only have we never seen it, there is no mysterious, unexplained, or unaccounted for phenomenon that can be ascribed to this missing particle. It was postulated purely because we trusted the integrity of the theory we had constructed. The renormalizability of electroweak theory, which makes it possible for us to extract sensible predictions from the equations, depends crucially on the existence of the Higgs field. We believe that the Higgs boson is out there, "waiting to be found," as 't Hooft puts it, but we will rest much easier once this "fugitive" is finally caught.

Setting a trap for a missing particle is an elaborate affair, and it is complicated further by the fact that we don't know where to look. While the Standard Model knows much, it remains silent on this matter; the mass of the Higgs boson is unknown. Until this particle turns up, colliders, present and future, will continue to create huge amounts of energy, over and over again, in the hope of causing a noticeable ripple in our invisible background and coaxing the Higgs field into emitting a boson. Our eyes are on Chicago's Fermilab and on the upcoming Large Hadron Collider: the state-of-the-art particle accelerator being built at CERN. In a decade or so from today, perhaps the Higgs boson will be spotted there, exactly where the W and Z were first found.

As his talk drew to a close, 't Hooft spoke about what comes next. With electroweak unification achieved, physicists couldn't resist the urge to go one step further. Could a Grand Unified Theory bring quantum chromodynamics—the theory of the strong nuclear force—into the fold? The mathematical details were worked out, and it appeared that the three quantum field theories might, at high enough energies, coalesce into one. In fact, if the world has one *more* hidden symmetry—supersymmetry, which relates forces and matter—grand unification would work very naturally.

How far can we push this scheme? Can we entwine the strand of gravity with this threefold braid, to end up with one "master" force? If that were possible, our four familiar fundamental forces would be like branches that grew out, at various times, from a single trunk. They would share common roots and belong to the same seed. Seductive though this prospect is, there are several hurdles along the way. Gravity cannot be described by a renormalizable quantum field theory, and so it stands apart for now. Will this obstacle ever be surmounted? Will we learn to fit gravity within our current paradigm, or will we need to stretch, or perhaps even reconfigure, our existing framework?

In recent years, a new candidate has risen, a theory that claims to achieve this ultimate goal, unifying all four forces into one, and including supersymmetry for good measure. The basic premise of this (super)string theory is that there is no such thing as a true point particle; instead, everything we see is made out of tiny vibrating strands of energy that are small enough to *appear* point like. String theory describes all known particles—matter and force alike—as modes of oscillation of these fundamental strings. Many particles can arise from the same fundamental string, just as a whole range of musical notes can be played on the same guitar string.

Could string theory be the holy grail of physics? Opinions are divided. No one can really say for sure yet. This morning, 't Hooft was certainly not ready to call it the final Theory of Everything. His penultimate slide showed a vibrating string, flipping over, turning itself inside out, performing all the deft maneuvers we would expect. String theory is aesthetically compelling, 't Hooft conceded, but it is not yet understood well enough for him to be convinced of its truth. There is still room for surprises, he said, as an image of magnifying glasses popped up on the screen behind him. If, 't Hooft continued, we could actually see a string, there's no telling what we would find. All of a sudden, music started playing, and the entire hall burst into laughter. As the magnifying glasses zoomed in, they revealed Taz—the Tasmanian devil from the Looney Tunes cartoons— jumping up and down on the string, using it as a tightrope, and pulling faces the entire time.

Most seminar speakers have to be reminded how much time they have left, but 't Hooft had built a very appropriate stopwatch into his presentation. Right at the bottom of the screen was a bundle of dynamite, tied up with a string lit on fire—a nod to well-loved cartoons as well as to Alfred Nobel, the inventor of dynamite and institutor of the prize. As the talk progressed, the spark came closer and closer to the bundle, until time ran out and the dynamite exploded over the screen. That is how it happens, 't Hooft concluded. Theories that contain paradoxes and inconsistencies eventually explode, leaving behind new and better theories in their stead. On this semi-philosophical, semi-humorous note, the lecture was over.

String theory intrigues me. I don't know if 't Hooft's criticism will be borne out, but I do agree with him that, throughout history, the devil has definitely been in the details. I never before pictured that metaphorical devil as Taz, but after today I will never be able to picture him otherwise.

Thinking back on it now, one of the reasons this talk stood out was 't Hooft's public affirmation that physics can be playful. It is liberating just to formulate those words in my mind. For years, I had secretly feared that I had too much fun with the ideas, that I did not treat them with the requisite solemnity, so I obviously wasn't "good enough." But 't Hooft testified, this morning, on an august stage, that you don't have to stand on ceremony with what you love.

I watch the sunset from the steps of the Opera House. At my feet, the Norrström flows like liquid paint. Way past spires and turrets, the Stadshuset tower is a speck in the distance. There, in the Blue Hall—which is all brick and no blue—the laureates will be awarded their medals as violet waves lap the walls and the Tre Kronor (three crowns) watch from atop the cupola in verdigris splendor.

Peace comes upon me, as it has not for a very long time. At the Aula Magna this morning, I had steeled myself for a painful visit to the past. I had gone to bid a final farewell to physics, and I expected to feel resentful. But my experience today had none of that drama or tragedy.

After the first tumultuous outpouring, doubt receded into the background. My regret at abandoning academics faded too. A profession is a way of earning a living, not a pledge of allegiance. Many people have hobbies and interests that define them far more than their jobs ever do. We use it that way, but the word *amateur* does not mean "unschooled"; it means one who pursues something for love, not to make a living.

Stockholm glows rose-gold in the dying rays of the sun. No matter where I live, this will always be my city. I have earned that right of ownership by enduring dark winters, taking long walks across cobbled streets, and discovering my own special, secret spots. Physics is my city too. I have rambled along her winding lanes and weathered her seasons. There are private rituals I enact, favorite places I revisit, every time I step back through her gates. Physics belongs to me by dint of labor and long years of love. No one can ever take that away. No matter where I may wander, physics will always be home.

The Final Chapter

From: Sara Byrne <breaking.symmetries@gmail.com>
Date: Fri, Mar 29, 2013 at 7:18 PM
Subject: Nature's Tapestry
To: Leonardo.Santorini@gmail.com

Dear Leo,

Thank you so much! The chocolates were delicious, and the card was lovely, but the last installment of your manuscript was, by far, my favorite birthday gift. It was devoured even faster than the truffles. Having pandered to your desire for instant gratification and immediate feedback, I have already gushed over these last two chapters on the phone, so I am not putting all that praise down in writing now. But I will share with you something else, and it *is* writing related . . .

A couple of days ago, I came across a little passage by Richard Feynman. I had meant to send it to you then, but it slipped my mind. Now it seems even more apt. "Nature uses only the longest threads to weave her patterns," Feynman wrote, "so each small piece of her fabric reveals the organization of the entire tapestry." And somehow that phrase reminded me of your book. I love the image of Nature sitting at her loom and weaving our world, but these words go deeper than that. The world is so rich and varied that at first glance it really does appear to be a patchwork of many different things, loosely tacked together. Yet every now and then, flashes of illumination expose a single glowing strand, meandering through the fabric, linking objects near and far. Isn't that exactly what happens when theories are unified? In a moment of revelation, a common underlying structure blazes into view.

As I listened to your narrators speak, I saw several motifs appear time and again, and my ears began to detect a rhythm in the repetition of their thoughts. Through the ages, and across the world, these very different men and women were intertwined by their shared loves and desires. There are some very long and beautiful threads in that human tapestry, too.

On a less lofty note, I did think about your proposal, and yes, I'll give it a shot. The way you spoke about it this morning, it actually sounded manageable. I'll just pretend I'm writing you a letter, and then you can cobble that randomness into a chapter that fits the book. I'm actually excited about it now . . .

More soon,

Sara

CHAPTER 7

The Fire in the Equations

[String Theory; Beyond the Standard Model]

WRITTEN BY SARA

*Even if there is only one possible unified
theory, it is just a set of rules and equations.
What is it that breathes fire into the equations
and makes a universe for them to describe?*
—STEPHEN HAWKING

APRIL 19, 2013
CAMBRIDGE, MA

Dear Leo,

It is a gorgeous spring day, and the afternoon light has outlined everything with a silver glow. In front of me, the Gothic arches of Memorial Hall are reflected in the long rectangular windows of the Science Center—a classic piece of art set off by a contemporary frame. Open-mouthed gargoyles dangle off the sunset-colored roof to stare down at all the people who run, bike, and walk across the plaza. Buildings are as empty as out-turned coat pockets, but every table and bench outdoors is packed to capacity. Sharp shadows splice the scene into polygons whose boundaries move, merge, or disappear under the shifting sun. I am amused by the faint relief I feel as shadows slide off objects that were split in two by their presence; it is soothing to watch the fractured patches of light and shade melt into a whole.

Amused, but not surprised. We have an innate urge to put together that which has been rent asunder; where this desire is fulfilled, we find beauty. Across disciplines, cultures, ages, and contexts, unity is considered to be "a fundamental—quite possibly the fundamental—aesthetic criterion."* I think this explains the public fascination with string theory. We are instinctively drawn to the possibility that everything we know might be reducible to a common origin; that the rich diversity of matter and forces in our universe reflects the range of motion of quivering, fluttering, infinitesimal strings. The idea has both intellectual and aesthetic appeal.

In almost every book or documentary that mentions this theory, the universe is likened to a string symphony. While not a precise mathematical statement, this makes for a pretty decent analogy. If we have to frame it in words, here's what we know: Fundamental strings, vibrating strands of energy, form the building blocks of nature. Strings can either be open—with two endpoints—or closed, like loops; all the particles we know, fermions and bosons alike, arise as the modes of oscillation of these strings. Just as with de Broglie's atomic orbits, the only oscillations that are sustainable are those whose wavelength divides the string into equal parts; all other oscillations interfere with themselves and decay. There are a discrete number of such possibilities,† corresponding to the distinct notes a string has in its repertoire. It takes a certain amount of energy to sustain each mode of oscillation, and from Einstein's mass-energy equivalence, it follows that every note has a characteristic mass.

Fundamental strings are truly miniscule. If a single atom became as large as the sun, a string would be approximately the size of a grain of sand. Strings lie so far beyond the resolving power of any microscope that their structure remains invisible; every string seems to be a mere dot, a point particle. But, of

---

*From the *Princeton Encyclopedia of Poetry and Poetics*.

†Corresponding to the fractions ½, ⅓, ¼, etc.

ONLY THE LONGEST THREADS

course, all strings are not alike, and though we cannot perceive the differences between them visually, we can tell them apart in other ways. Each oscillating string operates on a fixed energy, depending on the note it is playing at the time. The oscillations themselves escape detection, but we perceive their energy as mass. More frantic vibrations correspond to higher energy and, thus, heavier particles.

Unlike particle interactions, for which elaborate rules and explanations have to be concocted, strings interact in the most intuitive of ways: the ends of an open string join to form a closed string; or a closed string cuts open to form an open string; one closed string splits into two, or two closed strings combine to form one; one open string splits into two, or two open strings combine to form one. In this short list of moves is the root of all the games particles play. It is a truly elegant construction because so much arises, effortlessly, from so little. But the story isn't over yet. In the drama that is our universe, strings not only generate characters and write out their interactions, they also construct the background where it all plays out! This last, crucial step is string theory's major triumph.

Quantum field theory allowed us to incorporate matter and its interactions within a single framework. Symmetries held together particles that were susceptible to the same force. Of the four fundamental forces, quantum field theories were formulated for three; gravity remained the sole outlier. Once these forces were expressed in the same language, unification became possible. Electroweak theory combined the stories of electromagnetism and the weak force, and Grand Unified Theories were written down to include the strong force as well. But even though these expanded theories had a more ambitious scope and a larger symmetry, they were developed using the language and structure of quantum field theory, and that came with its own set of limitations.

Gravity was not resisting inclusion because it was stubborn, but because it was actively shut out by one of the foundational assumptions of this framework. Quantum field theory starts from the notion that space-time is a static stage on which interactions between particles play out. It concerns itself with describing the play. But we know, from general relativity, that space-time is dynamic and responds to the presence of matter. Before the four forces could be consolidated, a quantum theory of gravity would need to be formulated. It soon became clear that this would require some radical rethinking, and perhaps even the surrender of some dearly held beliefs.

A quantum description of gravity was desirable not just for abstract intellectual concerns like the lure of unification, but for more practical reasons. It does not often happen that the small and the heavy overlap, but when they do, the resulting systems and phenomena cannot be understood by appealing either to general relativity or to quantum mechanics alone. A reconciliation between the two is necessary for a sensible description of the early universe or of phenomena like black holes.

Popular accounts of string theory had just begun to flood the public consciousness when I entered my teens. On my thirteenth birthday, my parents bought me *The Elegant Universe*. "Calling it a cover-up would be far too dramatic," the book began. "But for more than half a century . . . physicists have been quietly aware of a dark cloud looming on a distant horizon." Straightaway I was hooked. No conflict between two characters could possibly be more epic than the battle between general relativity and quantum mechanics; no work of fiction could have a wider scope than string theory—the very fate of the universe hung in the balance!

From then on, I began to devour all the popular science accounts I could get my hands on. I knew that these were, of necessity, stripped-down versions of the truth, but I read on, and I proclaimed what I read. In retrospect, it gives me the shivers to think how obnoxious I must have sounded when I went around in high school saying, "The whole point of string theory is that there's no such thing as a point!" Awful. Pretentious. But kind of true; and also, in fact, the secret to string theory's success. Because strings had a finite extent, they did not squeeze and pinch quantities to zero size as point particles had done. Physical properties were no longer constrained and deformed into singularities; they finally had some breathing room. It wasn't much, of course, but still enough to keep them from screaming out mathematical expletives. To physicists who had become accustomed to swatting away infinities like flies, the renormalizability of string theory sounded like bliss.

What's more, string theory came with gravity built in. The four fundamental forces no longer needed to be understood separately and then merged somehow. There was nothing forced or artificial about this unification. The graviton, the force carrier of the gravitational field, was just another mode of oscillation of the string, as inevitable as any other. All the forces arose naturally from string theory; they all got equal billing. Space-time, finally released from the artificial frozen state imposed on it by quantum field theory, had a pulse again.

For its masterstroke, string theory broke down the wall between matter particles and force carriers. Strictly speaking, it is supersymmetry that is responsible for this final blow, which is why string theory is more properly called *super*string theory. Supersymmetry partners off bosons and fermions, letting each pair change identities through a symmetry transformation. The consequences are mind-blowing. If matter and force carriers can transform, each into the other, then they're both just different aspects of the same thing. It's like finding out Clark Kent and Superman are the same person; instantly, you know more about him than you ever knew about either of his two personas.

But for all it grants us, string theory demands something in return. It is a bit of a diva, actually, in the way it dictates its own terms and refuses to negotiate. The insistence on living in a ten-dimensional world, in particular, caused problems. Even among the usually elastic-minded tribe of physicists, many were unwilling to cede to this outrageous demand. Some were curious, willing to pay this price and see what ensued; others considered four dimensions to be sacrosanct and refused to enter the theater. But the fact remains that if you do cough up the price of admission, string theory puts on a dazzling show—like none you have ever seen before.

Over the years, our understanding has evolved so greatly that it has outgrown the theory's name. The ten-dimensional system of open and closed superstrings was just the theory's initial incarnation. Other objects lay hidden in the equations, too. Our suspicions were first raised when we saw open strings indulging in some rather strange behavior. Under certain conditions, these strings acted as if their motion was restrained—as if their endpoints were magnets, stuck to an invisible refrigerator door. Along the extent of the string, vibrations continued as before, and its ends appeared able to slide up and down and all across an invisible surface, but they were not, for some reason, free to move off the surface. We knew something funny was afoot, and when we investigated the situation further, we found that the strings were not play-acting; their motion was indeed confined, their ends stuck to hyperdimensional membranes we have come to call D-branes. These objects were not revealed earlier because they were not directly accessible to our scientific binoculars, which had focused only on the perturbations of strings. Only when this motion was affected in an inexplicable manner did we suspect the presence of something else. We changed tools, widened our focus, and suddenly D-branes came into view.

Such unexpected twists are part of the charm of theoretical exploration. To the extent that we cobble together its axioms, we play a role in the birth of a theory; but we can no more control its evolution or manipulate its final form than we can dictate the personality and fate of a child. There are always surprises. D-branes were not an unpleasant surprise, but there were others, and some came masked as crises.

When we had touted the "one ultimate theory" for a while, after we had brandished the banner of the ultimate unification and declared six hidden dimensions a paltry price to pay for such brilliance, we were hit by supreme embarrassment. String theory, it appeared, could be formulated in five perfectly consistent, perfectly legitimate versions. Imagine the irony! The final destination, where all roads were supposed to converge, was not a unique location. I found out about this conundrum years after it was resolved, and I still cringe when I think how awkward it must have been.

But a paradox is really just an opportunity to resolve a tricky tangle, and the solution of this particular problem led to a very profound insight. The five apparently distinct versions of string theory were like five different gates to the same city; they were connected, behind the scenes, via a web of dualities. The source of our embarrassment was that we had confused these entry points with the destination. This discovery raised the game to a whole new level. Remember the story of the blind men and the elephant? Five blind men each touch a different body part of the elephant, and their experiences are so different that each insists the others are lying. The conflict is resolved by the discovery that they all speak their own truth. There is no contradiction in multiplicity if one version does not have to be chosen above the others.

The five different ten-dimensional formulations led us to a single overarching theory, an eleven-dimensional beast we have named M-theory. We don't know a whole lot about it yet, other than this: the five superstring theories are the shadows it casts. Most objects are not perfectly symmetric, so when light hits from different angles, the shapes of their shadows change. These string theory shadows alone are not enough to construct M-Theory. They do tell us something, but there is obviously a lot left to explore. When we finally arrive at the terminal theory, it will probably look very different from the initial, just as an adult is sometimes unrecognizable from baby pictures, but I find it hard to believe that the beginnings are not there in string theory.

On another, more personal note, I think string theory gets unnecessarily bad press for being based on the "belief" of unification. Being a scientist means that you should be able to put your faith to the test, and be willing to surrender it should evidence so require. But it does not mean we have no beliefs. In fact, the very discipline of science is based on the belief that the world is comprehensible, and even more audaciously, that it is mathematically describable. The thing about scientific beliefs is that they are not arbitrary. They're not just random opinions we plucked out of the sky; we came to them through observation and experience, and so far, these beliefs have always been vindicated. It is so with unification as well. The history of physics is, in a way, a history of unification, of taking various apparently different strands and weaving them together. We have never had reason to question this pattern. And so, until it is contradicted, our assumption continues to be that even dangling, disconnected threads will eventually be incorporated into the fabric.

But of course we realize that this in itself is not an argument for the veracity of string theory. We recognize that nature is under no compulsion to model itself on our sensibilities, and that string theory will eventually need to make testable predictions. Contrary to what its critics contend, the reason such tests have not been conducted so far is not that string theory protests against experiment, or—like the ancient Greeks—does not want to sully itself by referring to the material world; it is simply that the theory is still developing.

It is much harder than before to test new ideas. We've pretty much exhausted the study of things we can hold. We could stop here and define physics as the study of all that is immediately accessible and directly verifiable; but if we follow our curiosity into realms that lie outside the natural grasp of our senses, we must be willing to extend our modes of analysis. We will have to invent ever more creative ways of probing these new frontiers,

ONLY THE LONGEST THREADS

often using indirect measurements, but that's the price of playing the game at this level.

It's a little like the story of Pinocchio,* actually. Among the many toys Geppetto has carved, the wooden puppet Pinocchio is his favorite. Geppetto is a good and gentle man who longs for a son, and the Blue Fairy grants this wish. She breathes life into Pinocchio and tells him that, if he handles the challenges of the world with grace, he will earn the right to become a "real boy."

Theoretical physicists are like Geppetto. We carve models out of mathematics. Some of our creations do not survive the blows of our own hammers and chisels, and others don't turn out as we had intended, but those that pass muster are sent out into the world with our blessings. If they prove their worth, they earn the right to be called theories; otherwise, they stay pretty mathematical toys forever.

Superstring theory is one of the most beautiful mathematical models we have ever made. If the Blue Fairy had to pick one of our creations to breathe life into, we can't help but think she'd choose this one. Perched on the workbench in our theoretical physics shed, it shows an uncanny knowledge of problems we didn't design it to solve. But, for all its promise, the fact remains that this puppet is only just learning to walk. We must wait a while longer for it to make concrete predictions about our four-dimensional world. For now, many physicists spend their days working on, and with, string theory. Some of them are right here, on the fourth floor of Jefferson Lab.

I remember how overwhelmed I felt when I walked up the stairs for my first official string theory lecture two years ago. Past the golden yellow wall and the bronze bust of Einstein were office doors that said: Andrew Strominger, Cumrun Vafa. I had heard the names mentioned reverently for years, and despite having seen these men at departmental gatherings, their reputations still dazzled me. In that moment, I felt inexpressibly

*The Disney version.

grateful for the chance to be initiated into the mysteries of string theory by its high priests.

Any fleeting fear that string theory might not live up to my expectations disappeared forever the moment that first class started. I felt a sense of homecoming. All those years of wrestling with atomic orbitals, statistical mechanics, circuits, and semiconductors paid off right then. Once I began to study the subject with the rigor and attention it deserves, my love for it grew even richer and deeper.

Sometimes, when I work out a problem, I feel a sense of pride in the way it unfolds on the page. "Look how clever this equation is!" I want to say. "Look at what it can do!" To ensure that I don't get swept up in adoration, I remind myself of Hawking's question: "Even if there is only one possible unified theory," he said, "it is just a set of rules and equations. What is it that breathes fire into the equations and makes a universe for them to describe?"

APRIL 25, 2013

Petals of magnolia and cherry blossoms rain down from the trees. I stepped in these pink puddles as I walked here to the bank of the Charles. I thought I would go over my chapter once more before sending it off to you, and I'm glad I waited. I now have the perfect ending.

Fabiola Gianotti visited the Physics Department this week to deliver one of the annual Loeb lectures. She spoke about the Higgs boson, of course. It's officially okay to call it that now. Intricate analyses have shown* that the particle which debuted last July is indeed a fundamental scalar that interacts with other particles in direct proportion to their mass; the boson is not Higgs-*like* anymore, she joked, even though we still like it.† She

---

*To one part in one trillion trillion.

†That drew a laugh from the crowd. The joke was not her own, Fabiola said, but she had to share it "because it was so cute."

showed us the graphs that led to these conclusions, saying she was "almost in tears" at their beauty.

The LHC had been running for three years before it was shut down for maintenance this February. It will be started up again in 2015. Before the beam was turned off, there was a push to collect as much data as possible. But with more data comes more noise; a lot of filtering out and sorting through will need to be done before we can unearth the gems that hide in this rubble. Some of these riches could, of course, be unexpected, but others could provide confirmation for some of our most cherished theories—supersymmetry being the prime contender and a favorite of many. Fabiola admits it makes her sad that we haven't seen the requisite evidence yet. "I love supersymmetry. It is a very canonical theory," she said. I couldn't help smiling. That's such a typically physicist form of praise! To us, that word has more than the standard English connotations. When a physicist says something is canonical, she does not just mean that it is authoritative, accepted, and worthy of inclusion in the canon; she also implies that it is the most natural and elegant choice to make under the circumstances. There is an undertone of logical inevitability, an implicit statement that it could not really have been otherwise, an acknowledgment of beauty.

Steven Weinberg expresses this feeling eloquently. "In listening to a piece of music or hearing a sonnet one sometimes feels an intense aesthetic pleasure at the sense that nothing in the work could be changed, that there is not one note or one word that you would want to have different," he writes. He explains why there is universal consensus among physicists that Einstein's theory of general relativity is more beautiful than Newtonian gravitation. The former is far more difficult to follow than the latter, but the ideas at its core are deeper and more basic. Newton's equations, while simpler in form, are not nearly as deeply rooted as Einstein's. Had experimental observations so required, Newton could easily have reshaped his equations without causing any offense, but once you start with Einstein's axioms, all roads lead to general relativity. And so, writes Weinberg, "Einstein's fourteen equations have an inevitability and hence beauty that Newton's three equations lack."*

After Fabiola admitted her love for supersymmetry, she went on to explain why she thinks this theory lies firmly within the realm of possibility. The observed mass of the Higgs boson is one indicator; the range in which this mass lies strengthens the likelihood that supersymmetry exists. According to current theoretical models, supersymmetry comes hand in hand with at least five Higgs bosons, but that isn't necessarily a problem. The one we have seen could very well be part of the quintet; or maybe we'll come up with modified models, who knows? Either way, this boson that popped up last July could provide a portal to a whole new realm of knowledge that was previously inaccessible.

One of the conjectures about dark matter is that it could be made up of superparticles. So, if supersymmetry is validated, we might be on our way to solving another mystery too. But even otherwise, supersymmetry is just a very cool theory. A few decades ago, we thought we were finally done tabulating particles. We'd split them up neatly into two categories, which obeyed different rules and had different physical purposes: fermions made up matter, and bosons carried forces. There was no possible way of confusing the two until supersymmetry came along, claiming that a boson could turn into a fermion, and vice versa, through a (super)symmetric transformation.

The most glaring issue with this theory is that we have not yet seen even one particle (boson or fermion) that could claim to be the superpartner of any other particle we know. Where are all these conjectured superpartners hiding? The answer lies in the fact that the universe around us today is not clearly supersymmetric. Bosons and fermions cannot switch identities anymore, merely by putting on, or taking off, a cape and glasses.

*Steven Weinberg, *Dreams of a Final Theory* (New York: Vintage Books, 1994), 132–36.

The conjecture is that supersymmetry was a feature of the early universe; as the universe evolved, it made asymmetric choices and, much like electroweak symmetry, supersymmetry too was broken. This mechanism,* we know, imparts wildly different masses to the particles affected. The claim is that supersymmetry's breaking made the superpartners so heavy that they are unable to exist at everyday energy scales. But the hope remains that the collisions at the LHC will give rise to some of them, at least the lighter ones. A proof of concept would be good to have.

A golden retriever runs along the track behind me, barking with the joy of being alive on a day like this. Parents push strollers along the path, pointing out the ducks to their toddlers. There are the usual cyclists and runners, and an occasional rowing crew passes by. Then there are people like me, content to bask in the toasty sunlight, floating in thought yet tethered to the ground by the exuberant sights and sounds of spring.

Like the winding river ahead, my thoughts veer also. But eventually they meander back to Fabiola's colloquium. Another wonderful discovery that may be lurking in the data is evidence for extra dimensions. Personally, I'm rooting for that to happen soon, so that people ease up on string theorists and our belief that the world has more dimensions than are apparent. By most accounts, even the largest of these dimensions are too small to be directly seen, but they should have observable, measurable consequences.

Despite its spectacular success, the Standard Model can't possibly be the final answer because several of its crucial features remain unexplained. It is like a page of text in which there are several blanks to be filled—like the mass of the Higgs boson, or the number of quark-lepton families—and no obvious way to fill them. Different choices lead to different universes. The ultimate theory, we expect, should be able to create everything from

*The masses of the W and Z bosons are very different from the mass of the photon. This discrepancy arises from the breaking of electroweak symmetry.

scratch; with just a few choice principles as input, our entire universe should emerge as the unique and inevitable output. A truly fundamental theory should make concrete predictions for every observable quantity. The purpose of experiments should be only to confirm these values, not to discover them.

The apparent scarcity of antimatter is another outstanding problem. The Standard Model makes room for antiparticles but doesn't say why we see so few. The natural assumption is that matter and antimatter should exist in equal amounts, but that is evidently not the case, at least in our corner of the universe. If it were, each particle would rush headlong into its antiparticle, and the pair would vanish in a burst of pure energy.

And then there is the fact that the Standard Model describes visible matter, which, we have recently learned, forms only 4 percent of our universe! We can't say much about the rest yet, except that it is dark, since it does not reflect light, and that some of it is matter, since it exerts a gravitational pull; the part that is not matter we call dark energy. Because they are invisible, these dark entities remained undetected for a very long time. The first clues about their existence came when astronomers observed certain galaxies rotating much faster than was justified by the amount of visible matter clustered around their centers. This would be possible only if the cores of these galaxies were much heavier than they appeared—the remaining matter, in other words, would have to be dark.* Astronomers track the effects of dark matter on cosmic scales, but it is up to particle physicists to discover its identity—to ask what, at the subatomic level, dark matter is actually made of. The study of space has long been separated from the study of particle physics, but to solve this problem, the two must come together again. There is a beautiful resonance between the large and the small.

The Standard Model has been checked countless times; there's no way it can be wrong. It will obviously need to be amended or

---

*Gravitational lensing is another phenomenon attributed to dark matter.

even superseded, but that will not make it incorrect. Like most theories, it has a limited domain of validity. I like the word *domain*. It reminds me of kingdoms. Each theory has its own kingdom to rule over; some are larger than others. We need to step beyond the borders of the Standard Model if we are to discover the identity of dark matter, understand dark energy, and solve the mystery of the matter-antimatter imbalance.

Einstein said, "No fairer destiny could be allotted to any physical theory, than that it should of itself point out the way to the introduction of a more comprehensive theory, in which it lives on as a limiting case." Science is a framework that evolves to fit our ever-expanding knowledge. We move forward by building on the theories of the past.

Less than ten months have passed since the Higgs particle burst onto the scene, but I feel I have lived through centuries. Perhaps it is the fact that your manuscript is done, perhaps it was listening to Fabiola again, or maybe it is a dawning understanding of the deep links between the minute and the colossal; whatever the reason, I cannot shake off the sense that, somewhere, threads are being tied together, and circles are closing in. Perhaps that is just how it works. The ring of knowledge breaks open every now and then to incorporate other strands and make the circumscribed area larger.

Once again, we have reached a bend in the road. As a scientist, I don't know if there is a more exciting feeling than this. It is extremely satisfying to understand phenomena; we are jubilant when experiments confirm our conjectures. But the best times of all are the moments of wonder and curiosity, when questions lie open in front of us, waiting to be explored, and we are aware that the resolution may be something we have never imagined in our wildest dreams. We stand again in that hallowed place where all manner of possibilities come alive. The prospects are dazzling.

Ciao,

Sara

Epilogue

From: Leonardo Santorini <leonardo.santorini@gmail.com>
Date: Tue, Oct 8, 2013 at 12:03 PM
Subject: Full Circle
To: breaking.symmetries@gmail.com

Dear Sara,

Congratulations! Our boson went from being a blip in the data to being lauded with the Nobel Prize! Now we can say we knew it when . . .

Also, I finally got through to my agent. She's called me over tomorrow morning to discuss her plans for the book. Things are happening so fast. I still haven't wrapped my head around all this. But I can only stay in New York till Sunday, and we now have several things to celebrate, so please get here soon.

Can't wait to see you.

Leo

Suggestions for Further Reading

Amir D. Aczel, *Present at the Creation*
Robyn Arianrhod, *Einstein's Heroes*
Jim Baggott, *The Quantum Story*
Bill Bryson (editor), *Seeing Further: The Story of Science, Discovery, and the Genius of the Royal Society*
Lewis Campbell & William Garnett, *The Life of James Clerk Maxwell*
Sean Carroll, *The Particle at the End of the Universe*
Frank Close, *The Infinity Puzzle*
Frank Close, Michael Marten, and Christine Sutton, *The Particle Odyssey*
Robert P. Crease, *The Great Equations*
Robert P. Crease and Charles C. Mann, *The Second Creation*
Edward Dolnick, *The Clockwork Universe*
Arthur Stanley Eddington, *The Nature of the Physical World*
Arthur Stanley Eddington, *Space, Time and Gravitation*
Albert Einstein, *Relativity: The Special and General Theory*
Albert Einstein and Leopold Infeld, *The Evolution of Physics*
Pedro G. Ferreira, *The Perfect Theory*
Richard Feynman, *QED: The Strange Theory of Light and Matter*
George Gamow, *Thirty Years that Shook Physics*
R. T. Glazebrook, *James Clerk Maxwell and Modern Physics*
James Gleick, *Genius*
James Gleick, *Isaac Newton*
Brian Greene, *The Elegant Universe*
John Gribbin, *The Scientists*
Paul Halpern, *Collider*
Paul Halpern, *The Great Beyond*
Sheilla Jones, *The Quantum Ten*
Leon Lederman, *Symmetry and the Beautiful Universe*
Joel Levy, *Newton's Notebook*
Basil Mahon, *The Man Who Changed Everything*
The New York Times, *A Search for the Higgs Boson*

Robert Oerter, *The Theory of Almost Everything*
Heinz Pagels, *The Cosmic Code*
Colin Pask, *Magnificent Principia*
Helen R. Quinn and Yossi Nir, *The Mystery of the Missing Antimatter*
Lisa Randall, *Higgs Discovery*
Edwin Emery Slosson, *Easy Lessons in Einstein*
Steven Weinberg, *Dreams of a Final Theory*
Frank Wilczek, *The Lightness of Being*
Shing-Tung Yau and Steve Nadis, *The Shape of Inner Space*

Acknowledgments

Without Bob and Ellen Kaplan, this book would not have been. Thank you, for seeing the potential in an early, incomplete draft, for being such enthusiastic advocates for the manuscript, and for your invaluable advice and counsel. I feel lucky to know you.

Working with Paul Dry Books has been an absolute pleasure. I am enormously grateful to Paul Dry for his guidance, support, and patience. My gratitude to Douglas Gordon for his thoughtful comments and meticulous attention to detail, and to Will Schofield for answering every question, large or small, with professionalism, friendliness, and promptness.

My heartfelt thanks to Freddy Cachazo, Ansar Fayyazuddin and Viqar Husain, who read with expert eyes and gave valuable feedback. I am indebted to Helen Quinn for pointing out several passages that might have been misleading, and to Katrin Schumann for her insightful comments. I am immensely grateful to Cumrun Vafa for the generous hospitality that is my first memory of Harvard. Thanks to the Royal Society for permission to use their library and to Aidan-Sean Randle Conde for answering questions about CERN. Fabiola Gianotti and Amir Aczel have been extremely gracious, and I am grateful for their encouragement. My thanks to Bilal Ahsan Malik and Ali Akram for the writing group that brought structure to my days, and to Rabia Saifullah, Maria Gatu Johnson, Maryanne Reilly, Catherine Mazur and Hans Hansson for their support.

At Grub Street, I found my writing home and met my writing family. I am particularly grateful to Ethan Gilsdorf, Mary Carroll Moore, and Tim Horvath, whose classes helped shape my book, and to Lynne Griffin and Katrin Schumann for all I learnt at the Launch Lab and for the very dear friends I made there. I owe a huge debt to the wonderful bookstores around me. Time and again, I found inspiration on the shelves at Harvard Book Store, the Coop, Porter Square Books, and Brookline Booksmith. Thanks to Harvard Book Store's amazing author series, I have had lovely conversations with writers I deeply admire and never expected to meet.

For the joy, humor, and perspective they bring, I am deeply grateful to Annie Durrani, Mehlaqa Samdani, Afia Nathaniel, Sameer Bajaj, Ayesha Tanzeem, Sidra Sheikh, and Aneeka Cheema—thank you for being the constants in my life. My in-laws have expressed pride and interest in this book right from the start. For their encouragement through the years, I am especially thankful to Fakhrunnisa and Fateh Khan and to Aisha, Saima, Mahvish, and Umer Malik. So much of what I do is due to the unwavering support and unending comfort I receive from my family. My parents, Basarat and Midhat Kazim, read every page of every draft, while Amir, Zaib, Mahe Zehra, and Ali Husain pretended to. I am grateful to them, for their love and patience, and to Asas and Murtaza, whose unflagging enthusiasm sustained me when my own wore thin. To Abdullah— thank you for being there every step of the way. You are my ideal reader.

ACKNOWLEDGMENTS